I0148797

SNIPER'S DISCRETION: UNVEILING AMERICA'S SHARPSHOOTER HISTORY

B. JAMES JOHNCOX

DEFIANCE PRESS & PUBLISHING, LLC

Sniper's Discretion: Unveiling America's Sharpshooter History

Copyright © 2024 B. James Johncox

(Defiance Press & Publishing, LLC)

Printed in the United States of America

10 9 8 7 6 5 4 3 2 1

All rights reserved. No part of this publication may be reproduced, distributed, or transmitted in any form or by any means, including photocopying, recording, or other electronic or mechanical methods, without the prior written permission of the publisher, except in the case of brief quotations embodied in critical reviews and certain other noncommercial uses permitted by copyright law.

DEFIANCE PRESS
& PUBLISHING

ISBN-978-1-963102-78-9 (Paperback)

ISBN-978-1-963102-77-2 (eBook)

Published by Defiance Press & Publishing, LLC, www.defiancepress.com

Bulk orders of this book may be obtained by contacting Defiance Press & Publishing, LLC., publishing@defiancepress.com

To my heroes - Lester, Jim, Mike, and Gregg
*who held me **up** through the many hard times.*
And
To my family, who made me feel like a hero myself.

CONTENTS

PROLOGUE

The history of the American marksman is a complicated adventure. Even the terms used to distinguish marksmen can be confusing. The words "marksman," "sharp-shooter," and "sniper" are seemingly used interchangeably and even among marksmen, the distinctions are not quite clear. From the moment that settlers first stepped foot onto the "New World," they had their work cut out for them. Agriculture was far from what it would become; thus the first settlers relied upon hunting and trade for survival. The firearm proved valuable for both ventures. It was valuable for hunting, for obvious reasons, but it was also valuable for trade with, and in many instances, protection against the various tribes and nations of Native Americans. By the end of the eighteenth century, colonial Americans would utilize it to prosecute a war of independence.

Technological advancements during this time frame would create a new type of firearm, the rifle, which would change the face of warfare. It increased the effective range of a fired projectile from 50 yards up to 300 yards, a 500 percent increase. By the mid-nineteenth century, further technological advancements, to include percussion cap ignition systems, aerodynamically superior projectiles, and the later development of metallic cartridge ammunition, combined with antiquated European warfare doctrines, resulted in the Civil War, the deadliest war Americans ever fought.

Nevertheless, it also produced a gifted class of marksmen, who were able to take shots at, and sometimes exceed, 1,000 yards. This was an astonishing feat even by modern standards. These gifted marksmen, previously known as "riflemen," were given the title of "Sharpshooter." Outside of warfare, this inevitably resulted in contests of skill whereby marksmen demonstrated their prowess with the rifle to gathered spectators and/or had their feats celebrated and published in the newspapers. Contests of this nature still take place at present in the form of Palma matches, F-Class rifle matches, National Rifle League matches, and the Precision Rifle Series to name a few.

By the beginning of the twentieth century, the marksmanship skillset remained the same. Within the military, however, telescopic optics, field craft training, and observation skills were added. Their title, thanks to their British allies, was changed to "sniper." It would seem as though no nation since has so completely endorsed that which could be considered a martial skill. However, this is simply not the case. While competitive feats of precision marksmanship are celebrated, the sharpshooter/sniper has, until recently, been condemned within American culture. Even the word "sniper" is used liberally to describe an individual who commits murder via use of any rifle at any distance. The truth, unfortunately, is that throughout American military history, the United States has abandoned its sharpshooter/sniper programs after each major conflict up to, and including, the Vietnam War. While many historians acknowledge this, few speculate as to why these programs were abandoned, and none addressed the long-term implications it would have on military preparedness.

This research seeks to shed light on this dilemma via an in-depth look at sharpshooters/snipers and the programs with which they were affiliated in each of these major conflicts in American military history. Suffice it to say, whether one is comforted by the idea or finds it to be anathema, American culture is inextricably tied to the rifle, and by extension, the marksmen who wield them.

PART I

Early American Warfare

"When shoes and clothes and food, when hope is gone we'll all have the rifle."

- John Steinbeck, *The Grapes of Wrath*

CHAPTER I

THE AMERICAN REVOLUTION

Many, when considering the American War of Independence, otherwise known as the American Revolution, imagine non-uniformed, untrained minute men firing muskets at red clad British regulars from behind rocks, trees, and bushes. This characterization is iconic and, to many Americans—heroic. Unfortunately, it is also largely myth. That is not to say that this type of guerrilla tactic did not occur; however, the vast majority of battles took place utilizing the traditional European warfare doctrine, whereby troops stood in rank and file and marched towards one another until within musket range. Upon halting, the combatants would exchange volleys of musket fire. If one of the phalanx formations showed signs of breaking, a bayonet charge would ensue to force a rout and the army that occupied the field when the fighting ceased was declared the victor. This begs the question, "Where did this myth of firing from behind rocks and trees come from?"

In 1775, when the first battles of the American Revolution began, rifles found their way onto the battlefield. Although traditional European methods of linear warfare dictated that smoothbore muskets fitted with a bayonet were the weapon of choice for war, the Continental Congress had made a provision within the June 14, 1775 act that raised the Continental Army. The provision stated "six companies of expert rifflemen [sic], be immediately raised in Pensylvania [sic], two in Maryland, and two in Virginia..."[1] Before one can understand how significant this provision of the Continental Congress was, a distinction must be drawn between the rifle and its predecessor, the smooth-bore musket. The word "rifle" was not used interchangeably with the word "musket." The smooth-bore musket, as the name entails, possessed a barrel with a smooth interior, or bore.

When a projectile was fired from the musket, it left the muzzle somewhat haphazardly. This limited the effective range of a smooth-bore musket to roughly 50 yards. In contrast, the rifled musket, or "rifle," possessed a bore that featured lands and grooves, or "rifling," that ran the length of the bore in a spiraled pattern. When a projectile was fired from the rifle, it left the muzzle with a spin that created gyroscopic stabilization in flight. This stabilized flight allowed the spherical lead balls to be accurate to approximately 300 yards. Furthermore, a standing army was not referred to as "riflemen," let alone "expert riflemen," it was called an army. Thus, it stands to reason that the Continental Congress chose its phrasing carefully. They wanted ten companies of expert riflemen, or 800 soldiers capable of engaging enemy soldiers at distances beyond the range of a musket.[2] The United States Army of today credits this provision with the birth of the American Army. Although this provision does suffice in this capacity, it could, however, be argued that the creation of the U.S. Army was, in fact, the creation of the world's first elite military sharpshooters. The next paragraph in the act states that these companies of expert riflemen "shall march and join the army near Boston, to be there employed as light infantry..."[3] This leads one to believe that a standing army of infantry already existed and that the Congress was establishing these riflemen to supplement the ranks. Although the army near Boston was comprised of militia, and the Congress was, in effect, creating the first *federal* fighting force, the language used was quite particular. As opposed to establishing a *continental army,* the delegates chose instead to create an army of expert riflemen. Indeed, they were to be the first force multiplier echelon in American military history. This is further corroborated by John Adams in a letter to Elbridge Gerry, who at the time was serving in the Massachusetts House of Representatives and would later serve

---

1. Worthington C. Ford, ed., *Journals of the Continental Congress: 1774-1789*, vol. 2 (Washington: Government Print Office, 1904), 89.

2. Worthington C. Ford, *Journals of the Continental Congress*, 89.

3. Ibid., 89

in many other political roles, to include Vice President. The letter, dated June 18, 1775, reads:

> "... ten companies of riflemen be sent immediately; six from Pennsylvania,
> two from Maryland, and two from Virginia, consisting of sixty-eight
> privates in each company, to join our army at Boston. These are said to be
> all exquisite marksmen, and by means of the excellence of their firelocks,
> as well as their skill in the use of them, to send sure destruction to great
> distances."[4]

Thus, according to the language utilized in the *Journals of the Continental Congress* and in John Adams' letter to Mr. Gerry, compared to the tactics and vernacular of that era, this resolution, in fact, was particularly aimed at founding the first sniper units in world history.

This is a bold, if not . . . a provocative, claim indeed. Nevertheless, when conducting an investigation, one must follow the evidence despite the outcome. The Battle of Bunker Hill began three days after this resolution was passed. It could be argued that these congressmen saw that the fledgling nation was standing at the precipice of war with the most powerful empire the world had ever seen and recognized that a relatively small force of men, equipped with technologically advanced weapons, could turn the tide of battles. Simply stated, they needed force multipliers. It proved to be exactly what the colonies needed at exactly the right time. This was not by accident. The importance of a written record cannot be overstated. After all, why did the nation's founders decide to create a document, albeit brief, on their discussions with one another? Perhaps it was to establish a written account of representation . . . an effort to protect the rights of the people that had been trounced upon by the British crown, or maybe it was foresight . . . manuscripts created for posterity, so to speak. These reasons may be partially true; however, the less elegant, and far more likely answer is that they needed a method by which to remember the myriad of issues they were addressing and the methods by which they solved problems. Insofar as rights are concerned, keep in mind that the Declaration of Independence would not be penned for another year, and the Constitution would not even be considered for

---

4. John Adams, "From John Adams to Elbridge Gerry," *The John Adams Papers*, ed. Robert J Taylor (Cambridge, MA: Harvard University Press, 1979), 25-27.

another decade. Despite the justification for keeping the record, language was used far more carefully then than it is now. A written message, at that time, carried with it far more weight and value than it does today. At that time, errors in written language were not acceptable. The misuse or misunderstanding of vocabulary could have brought with it the demise of what would become the United States of America.

In this case, the most important piece of evidence boils down to two words, "expert riflemen." The argument might be made that the Congress was speaking in generalities; however, that would not have been the case. Although the word "rifle" is commonly used today, in 1775, a rifle was a state-of-the-art advancement in firearms technology and was absolutely incongruent with the European, Napoleonic, linear warfare doctrine—the accepted method of warfare during that era. In addition to this, troops were not referred to as riflemen. Then, similar to now, troops en masse were referred to as troops, soldiers, regulars, militia, and armies. Why then, in one singular document, would the Continental Congress refer to this particular group of standard soldiers as riflemen? The simple answer is they would not, nor did they. They documented exactly that which they desired—namely, men who were skilled with a firearm that outmatched the musket in accuracy. Unfortunately, the transcription in the journal is brief. The entirety of the day's events makes up only two pages of the journal, thus exactly what was discussed and stated will almost certainly never be fully known. Indeed, the only true clarification is contained in Adam's letter to Gerry where he notes the soldier's expertise and the "great distance" at which the troops will be effective.

Notwithstanding, it is a certainty that the standard military issue firearm of that era was the musket, specifically the British Land Pattern musket, commonly referred to as the "Brown Bess," and that the musket was seldom confused for a rifle. Rifles would have been poorly suited to warfare and were more often found in the possession of hunters and trappers. The smoothbore musket, on the other hand, could be fired and reloaded four times a minute by a well-trained soldier. The accuracy of a rifle was predicated on the projectile fitting tightly against the lands and grooves of the bore, greatly increasing reloading time and greatly reducing the volume of fire that could be generated by a fighting formation. Therefore, though the rifle existed in the late eighteenth century, it was not considered a suitable weapon for linear combat.

Non-linear combat; however, is another story altogether. Throughout the Revolutionary War, Americans did not exclusively use guerrilla tactics; however, the use of guerrilla tactics was almost always exclusively American. Indeed, it is often believed that the only way that the British, then the strongest military in the world, could have been defeated is via the widespread use of guerilla tactics, what the British derisively called the "American Way" of war. Reverence is given to the idea of the outgunned and outmatched under-dog colonists, who were firing rifles from behind rocks, trees, and bushes. While that did happen, at the very first battles at Lexington and Concord for instance, it was extremely rare. That is not to say that rifles were not used during linear battles. Riflemen were often used as skirmishers and, more effectively, as what would today be recognized as a "sniper." Some of these men served under Colonel Daniel Morgan and came to be known as Morgan's Riflemen, one of whom can be credited with ending the British counterattack at the Battle of Bemis Heights by targeting and killing British General Simon Fraser at approximately one quarter of a mile, or roughly 400 yards.[5]

Notwithstanding, the American Continental Army was still much smaller than their British counterparts, and the use of local American militias to supplement the ranks was common. Typically, militia units were not uniformed, and often the individual soldiers were armed with the personally owned firearm that they brought with them. Although these militia units were, in most cases, armed with smoothbore muskets, some of the militiamen carried rifles of their own. When the British encountered these riflemen and militias, they found themselves facing a new enemy. Specifically, these were men who had carved out a living in the Appalachian wilderness as hunters and trappers. Some, inspired by the encroachment upon their land during the British Southern Strategy, marched over the Appalachian Mountains to meet the British in battle. These men became known as "Overmountain Men." This was a result of necessity.

American forces, especially in the south, were comprised almost entirely of militias who did not have the numbers to stand in linear formations against the much larger British Regular forces. Thus, the Patriot militias became resourceful, utilizing their knowledge of the frontier, their own clothing and their own hunting rifles. Their unique frontiersman appearance and tactics were developed not for war, but for survival in the Appalachian Mountains of Virginia and Tennessee. One such example is known as the Battle of King's Mountain. On October 7, 1780, the Battle of Kings Mountain took place on the northern border of South Carolina. One characteristic of this battle that stands

---

5. William L. Stone, *Campaign of Lieut. Gen. John Burgoyne: and the Expedition of Lieut. Col. Barry St. Leger* (Albany, NY, 1877), 62.

out above the others is that the Battle of Kings Mountain was fought almost entirely by Americans. That is to say, American militias comprised the ranks of both sides.

The British Tories were led by Major Patrick Ferguson, who British General Charles Lord Cornwallis had installed as the British Inspector of Militia.[6] The American Patriots were led by Colonel William Campbell and hailed mostly from Virginia and what is now known as Tennessee.[7] The topography of Kings Mountain was ideal for the Patriots' tactics, as the slopes were densely forested and had large trees and boulders to take cover behind. This made forming a linear battle formation impossible for the Patriots assaulting up the mountain and provided difficulty for the British Tories, who controlled the treeless top.[8] Nevertheless, instead of building and maintaining ramparts or defensive fortifications at the top of Kings Mountain, Ferguson ordered his men to perform a bayonet charge down the hill, towards the assaulting Patriots.

Perhaps one of the most defining characteristics of this particular battle was the use of rifles. On Kings Mountain, many of the Patriots were armed with their own rifles. The Patriots, having utilized their rifles to hunt as a means of survival, were quite adept at delivering extremely accurate fire at distances exceeding the capabilities of smoothbore muskets. Each time the Tories attempted to repel the Patriot assault up the mountain, the patriots would simply peel off to the flanks and continue accurate fire.[9] Although the Tories were also American, they had been trained by the British Army to fight in the traditional European method and were unaccustomed to the tactics utilized by Campbell's "Overmountain Men."

The battle lasted approximately 65 minutes and at the end, 290 Tories were killed including Major Ferguson; 163 were wounded; and 668 were missing or captured. Patriot casualties numbered just 28 killed and 62 wounded.[10] The traditional European method of battle had utterly failed against the Patriot's guerrilla tactics and accurate rifle fire. What is of particular interest in this battle is that Major Patrick Ferguson, the British commander, had been instrumental and successful in establishing marksmen within the

---

6. "Battle of Kings Mountain Facts & Summary," American Battlefield Trust, January 14, 2020,

7. Ibid.

8. "Battle of King's Mountain," February 9, 2010, https://www.history.com/this-day-in-history/battle-of-kings -mountain .

9. "Battle of Kings Mountain Facts & Summary," American Battlefield Trust, January 14, 2020.

10. "Battle of Kings Mountain Facts & Summary," American Battlefield Trust, January 14, 2020.

British ranks and had designed the Ferguson Rifle. Indeed, earlier in the war, at the Battle of Brandywine, Ferguson, who was considered the finest marksman in the British Army, was armed with the rifle that bore his own name. He had taken careful aim at an American officer with a large hat. Although he had a clear shot and plenty of time, he ultimately chose not to fire. His code of chivalry dictated that it was dishonorable to shoot in the back an enemy officer who was not engaged in combat. The officer with the large hat he aimed at was none other than General George Washington.

Nevertheless, the results of the battle at King's Mountain forced Lord Cornwallis to remove his remaining troops in the northern colonies and focus them entirely on his Southern Campaign. His Commander-in-Chief, Sir Henry Clinton, stated of the battle that it was "the first link in a chain of evils that followed each other in regular succession until they at last ended in the total loss of America."[11] Armed with rifles and wearing buckskin clothing that blended with the environment, these men utilized techniques in battle that were cultivated over a lifetime of hunting in the wilderness and fighting Native Americans. The British were altogether unprepared to meet this "American Way" of war and regarded it with contempt, deeming it dishonorable.

Although there exists more anecdotal evidence of riflemen participating in battles during the course of the American Revolution, it may be superfluous to include them in this text. That is not to say they are not important when considering the broader context of the American Revolutionary War. This is, after all, meant to be an historical text. However, considering the regard with which these marksmen were held, their deeds were seldom documented. In those events when it was, there is little more than speculation as to the identity of the individual who prosecuted the shot. In the instance of General Fraser's death at Bemis Heights, it is widely believed that an individual named Timothy Murphy fired the shot; however, there exists no empirical evidence to confirm or deny this.

Nevertheless, each feat of marksmanship has common denominators. Namely, one man . . . one rifle . . . one target. It is this maxim that has ascribed, at least until recently, the malevolent disposition and contempt with which snipers/sharpshooters were held. Indeed, it seems as though individuals who discriminately chose targets while engaged in combat, were somehow considered unscrupulous; whereas those who fired indiscriminately—cannister or grape shot for instance—were worthy of reprieve.

---

11. George C. Mackenzie, *Kings Mountain* (Washington: National Park Service, 1961), 1.

Subsequent to the Treaty of Paris being signed and formally ending the American War of Independence in 1783, General George Washington disbanded a large portion of the Continental Army. To say that the companies of expert riflemen created by the Continental Congress were disbanded due to the contempt with which they were held would be inaccurate. Robert Wright points out in his book *The Continental Army*, that "Congress adopted a general resolution on 23 April that was a compromise between those members who wished a swift disbandment to reduce expenses and those who were hesitant to act until the British had evacuated their last posts."[12]

Although riflemen were not singled out to be abolished, the cost of maintaining a large army was too great for a fledgling nation. With regards to what would remain as a national fighting force, Washington requested four infantry regiments assigned to the western frontier, a national military academy to train artillery and engineering officers, and one artillery regiment.[13] No specific mention of riflemen was made. This is most likely due to Washington's own military training and experience. Although he was named Commander in Chief of the first American Continental Army, he was, after all, a British trained military officer whose own proclivities gravitated towards the traditional warfare doctrine of the age. Thus, the expert riflemen of the American Revolution, due in large part to cost prohibition, quietly returned to their homes and were largely forgotten.

---

12. Robert K. Wright, *The Continental Army* (Washington, D.C.: Center of Military History, U.S. Army, 1989), 179.

13. Ibid.

NOTES:

1. Worthington C. Ford, ed., *Journals of the Continental Congress: 1774-1789*, vol. 2 (Washington: Government Print Office, 1904), 89.

2. Worthington C. Ford, *Journals of the Continental Congress,* 89.

3. Ibid., 89

4. John Adams, "From John Adams to Elbridge Gerry," *The John Adams Papers*, ed. Robert J Taylor (Cambridge, MA: Harvard University Press, 1979), 25-27.

5. William L. Stone, *Campaign of Lieut. Gen. John Burgoyne: and the Expedition of Lieut. Col. Barry St. Leger* (Albany, NY, 1877), 62.

6. "Battle of Kings Mountain Facts & Summary," American Battlefield Trust, January 14, 2020,

7. Ibid.

8. "Battle of King's Mountain," February 9, 2010, https://www.history.com/thi s-day-in-history/battle-of-kings-mountain.

9. "Battle of Kings Mountain Facts & Summary," American Battlefield Trust, January 14, 2020,

10. Ibid.

11. George C. Mackenzie, *Kings Mountain* (Washington: National Park Service,

1961), 1.

12. Robert K. Wright, *The Continental Army* (Washington, DC: Center of Military History, U.S. Army, 1989), 179.

13. Ibid., 180.

CHAPTER II

THE AMERICAN CIVIL WAR

At the outbreak of the American Civil War in 1861, much had changed in the United States, that is much except battlefield tactics. As armies of the North and the South met each other in battle, the traditional warfare doctrine that was used almost a century earlier, during the American War for Independence, was once again utilized. Unfortunately, there existed a blueprint of what was to follow.

The Crimean War (1853-1856) that took place only a few short years prior to the outbreak of the American Civil War was a grim foreshadowing of the carnage that would be wrought on American troops. The Crimea, a peninsula located in the Black Sea, was a Russian territory at the time. Though control of the peninsula has changed hands a number of times since, as of the penning of this text, it is under the de facto control of Russia. Nevertheless, the Crimean War erupted over religious disputes between the Catholic, Eastern Orthodox, and Ottoman Empire churches. Notwithstanding, cause is not the subject of consideration for the purposes of this text, but the devastation manifested by the combination of technological advancement and archaic tactics. Indeed, it demonstrated that the combination of a reliable ignition system, a rifled bore, and a superior projectile coupled with antiquated tactics were the ingredients necessary to deliver death wholesale. The recipe was so volatile that the Crimean War's lasting legacy includes producing historical titans such as Florence Nightingale and Leo Tolstoy.

Martin Pegler, the former Senior Curator of the Royal Armouries Museum in Leeds, England points out in his text *Sniper: A History of the US Marksman*: "The use of rifles quickly turned this accepted [linear warfare] military tactic on its head. Defenders could open accurate fire at 500 yards or more and decimate attackers before they could get anywhere near a range at which they could pause and return fire."[1] Indeed, a number of inventions had exponentially increased the lethality of a martial force. Although some of the firearms on the battlefields of the Civil War were still smoothbore muskets, rifles had seen an increase in production and use.

Furthermore, flintlock firearms had been replaced by percussion cap systems, also known as "caplocks," which were far more reliable. The spherical lead ball projectiles of the past had given way to a more aerodynamic projectile initially designed by a French military officer, Claude-Étienne Minié. The minie ball, as it would come to be known, was a conical shape lead projectile with a concave base. This design allowed for the bullet to be slightly smaller than the bore from which it would be fired, making the loading of rifles easier. Prior to this, the loading and reloading of rifles, in comparison to smooth bore muskets, was a far more time-consuming process as the projectile had to fit snug against the rifling to ensure gyroscopic spin was generated upon ignition. The addition of the concave base of the projectile meant that, upon ignition, the expanding gasses would cause the cup at the base of the bullet to expand, forcing it into contact with the lands and grooves of the rifling.

On the other side of the world, throughout the Antebellum period of America, tensions grew over the institution of slavery. After many failed compromises and the infamous Dredd Scott Supreme Court decision, it seemed as though hostilities would inevitably erupt. This tension came to a boiling point when, in December 1860, after the Presidential election of Abraham Lincoln and during President James Buchanan's "lame-duck" period, South Carolina seceded from the Union. Six more states would follow suit before Lincoln ever took office. On April 12, 1861, the newly formed Confederate States of America under President Jefferson Davis launched a preemptive strike on Fort Sumter in Charleston Harbor in Charleston, South Carolina. The attack consisted of twenty-four hours of artillery bombardment followed by the surrender of the Union forces that held the fort. The casualties generated by the attack numbered zero; however, this would not be an indication of what was to come.

---

1. Martin Pegler, *Sniper: A History of the US Marksman* (London: Osprey Publishing, 2011), 71.

Although the rifle had become more prevalent on the battlefield, a number of commanders understood its capabilities better than their counterparts. One such commander was a Northern gentleman named Hiram Berdan. Berdan was a wealthy businessman and a long-range precision marksmanship enthusiast. Pegler wrote of him that he was "an extremely clever engineer [but] he had some serious character flaws..." and that he was a "supreme egoist and an obsessive self-publicist, and he appeared on many occasions extremely reluctant to actually place himself in any physical danger."[2] While the first of those characteristics is generally not considered *ideal* for a military sharpshooter/sniper, the latter is not necessarily a poor quality. It is arguable that Berdan had little common interest in abolitionist causes, and that he was more concerned with his own legacy. Given that he had the capital and political connections necessary to do so, on June 15, 1861, Berdan was given the authority to raise a regiment of specialized marksmen for the Union Army.[3] Despite his eccentricities or motivations, Berdan was a gifted marksman and knew what would be required of his regiment. He placed advertisements in newspapers in order to recruit men for what would become the United States Sharpshooters (USSS). One such ad not only displayed Berdan's hubris, but it also made clear the skill which was required to be counted among his ranks:

THE SHARPSHOOTERS OF THE NORTH.

Fifth Avenue Hotel, New York, May 30, 1861.

To the Sharpshooters in the Loyal States:-

Gentlemen – Many of you are undoubtedly aware that an effort is making
to get up a regiment to be composed entirely of first class rifle shots at
long distances, and that in consequence of my having done something
in this way of rifle shooting, suggestions have been made in the public
press that I should aid in the effort. I am, moreover, receiving almost daily
applications, by letter and in person, to the same effect; and I see so clearly
the great importance of the object in view, that I do not feel at liberty to
refrain from doing what I can to further it. With this view I propose that
all gentlemen who have made themselves good shots at long distance, who
are willing to place their skill in this way at the service of our country in

---

2. Pegler, *Sniper: A History of the US Marksman*, 72.

3. Pegler, *Sniper: A History of the US Marksman*, 72.

this her great struggle, should send their names to me, with an affidavit showing the best shooting they have done at two hundred yards, or more. As soon as the necessary arrangements are made for equipments [sic], &c., notice will be given to all those whose applications are approved. No application will be considered in which the average of ten consecutive shots exceeds five inches from the centre [sic] of the target to the centre [sic] of the ball at two hundred yards. The prodigious efficiency of detachments of such sharpshooters, armed with our northern patent target rifles, needs only to be alluded to be recognized at once by all who have any knowledge of the subject. Need I add one word to enforce the duty of our amateur target shots to make their peculiar skill useful to our country at this time of trial? That skill – the offspring of a manly Northern sport – can be converted into a powerful military instrument so readily, I feel confident the subject need only be suggested to insure its being fully and promptly attended to. Very respectfully, your obedient servant.   H. Berdan[4]

Berdan obtained a commission as a Colonel and created the 1st and 2nd United States Sharpshooters (USSS) regiments. A popular myth surrounding the title "Sharpshooter" is that it came as a result of Christian Sharp's design of rifles that the USSS were later issued. It would be convenient if that were the case; however, most never utilized the Sharps rifle, and although it did eventually become the standard issue, those who did not furnish their own target rifles, some of which weighed in excess of thirty pounds, were initially issued with a Springfield M1861.[5] The truth is that the word sharpshooter, or *scharfschütze,* had already been in use since the seventeenth century "where it was commonly used in Germany and Switzerland to describe a good shot."[6] The word *scharfschütze* is still utilized in German speaking countries today to denote a sniper qualification.

Berdan's 1st and 2nd United States Sharpshooters would go on to participate in many of the battles throughout the American Civil War. Perhaps their most distinguished acts took place at the Battle of Gettysburg where they were instrumental in defending the various hilltops from Confederate charges. The sharpshooters were easily distinguished from their Union counterparts. Instead of the blue uniforms that were standard issue to

---

5. Pegler, *Sniper: A History of the US Marksman,* 77.

6. Ibid., 72.

Union soldiers, Berdan had requested a uniform specific to the USSS. Captain Charles A. Stevens, who served under Berdan, described the uniform as "consisting of dark green coat and cap with black plume, light blue trowsers [sic] (afterwards exchanged for green ones) and leather leggins [sic]..."[7] the notion being that the colors helped in the concealment of his men among the foliage.

This contrast; however, had a negative impact insofar as social implications. The USSS considered themselves to be elite among the ranks and in many respects they were. Unfortunately, this earned them extensive criticism and disdain from their non-sharpshooter comrades in arms with one sharpshooter commenting that it seemed as though they were "hated by all that have to deal with us."[8] Furthermore, throughout the war, sharpshooters on both sides became increasingly more despised. Sharpshooters who were caught by enemy soldiers would often hide their rifles to avoid being identified as such to prevent a summary execution or being hanged as a sharpshooter.[9] This is perhaps best illustrated by a recruitment article that states, "The design of the Colonel is to have the regiment detached in squads on the field of battle to do duty in picking off officers and gunners on the European plan, by which they take the risk of being cut off by cavalry, or executed, as they certainly would be if taken."[10]

Insofar as the issued rifles themselves are concerned, Berdan had his work cut out for him. After all, establishing a unit of sharpshooters is entirely predicated on providing said sharpshooters with a suitable rifle capable of the accuracy required. Much then as it is now, few understood the intricacies involved in accuracy and precision. Today, most would not know the difference between a sniper weapon system and a typical hunting rifle with the exception that most "deer" rifles have a wood stock.

Then, like it is now, the number one requirement was a heavy barrel. More often than not, this meant that Berdan's men supplied their own rifles . . . many of which were crafted by individual gunsmiths and varied in caliber from rifle to rifle. Some of these handcrafted rifles weighed upwards of thirty pounds or more. This brought with it a number of problems. Heavy rifles are cumbersome and are not exactly well suited for

---

7. Charles Augustus Stevens, *Berdan's United States Sharpshooters in the Army of the Potomac, 1861-1865*, reprint, (Bethesda, MD: University Publications of America, 1994), 10.

8. Caspar Trepp, "Letters from Captain Isler," *Papers*, accessed January 6, 2021.

9. Pegler, *Sniper: A History of the US Marksman*, 88.

10. "Col. Berdan's Sharpshooters," *New York Times*, August 7, 1861, p. 8.

war. However, the more pressing logistical problem was the varying calibers. Although breech loaded firearms, as well as self-contained metallic cartridges, would start to make appearances during the American Civil War, most firearms, rifled and smoothbore alike, were still loaded from the muzzle. This may conjure images of men utilizing a powder horn to charge their rifle, and that may have occasionally been the case, however, pre-rolled paper cartridges were the standard. The paper cartridge was not what one would imagine a rifle cartridge to be today. Although it was self-contained, complete with the correct grain weight of powder, a projectile, and the paper itself that could be utilized as the wadding, it was not loaded into the chamber of the rifle like a metallic cartridge is today. Instead, the paper would be ripped; the powder dumped down the barrel; the ball (i.e. minie ball) would be placed atop a patch of the paper and subsequently rammed down the barrel with a ramrod. Unlike today, there were no organizations that maintained a list of standardized projectile calibers; thus some sharpshooters had .58 caliber rifles while others had .69 caliber etc. . . . Berdan sought to remedy this in the same way that any competent military commander would, by issuing standardized equipment to his men.

Nevertheless, Berdan would experience friction in accomplishing this task. Most soldiers, both Union and Confederate, were issued the Springfield Model 1861 rifle in .58 caliber. Although these rifles were capable of accurate fire, Berdan sought better. Specifically, he wanted to issue his men with Christian Sharps Model 1859 Improved Rifle. The Sharps New Model 1859 was equipped with a double-set trigger and a modified breech lever that also served as the trigger guard. The Sharps fired a .52 caliber projectile that was prepackaged in a linen cartridge, which significantly expedited loading. Most importantly, the Sharps was extremely accurate, even by modern standards. In the hands of a skilled marksman, it was capable of engaging targets in excess of 1,000 yards. Unfortunately, it was also extremely expensive. Prohibitively so. Each Sharps rifle would cost the Union approximately $45.00 (roughly $1,421.00 as of 2024.) The War Department, and specifically the Army Chief of Ordnance General James Wolfe Ripley, believed this to be outside the realm of possibility. Notwithstanding, Berdan was in regular contact with powerful individuals, to include Major General George McClellan, who was in charge of the Army of the Potomac and the Secretary of War Simon Cameron.

Throughout the ordeal Berdan tested other rifles from other manufacturers as well, to include a submission from the Colt company. The Colt Army Rifled Musket featured a fluted five chamber cylinder, a 37 ½ inch barrel, and was chambered in .56 caliber. Although it was not the preferred offering, Berdan approved of its performance; thus

Berdan entered into negotiations to obtain the Colt as well. Despite repeated pleas on Berdan's behalf from McClellan and Cameron, Ripley continuously refused to place the order for Berdan's rifles, Sharps or Colts . . . that is until President Lincoln was made aware of the ongoing consternation and directly ordered Ripley to order the rifles. Berdan, as stated previously, was a gifted businessman.

In the end, much to his chagrin, Ripley ordered both the Colt and the Sharps rifles. Although Berdan had promised his men the Sharps, it was the Colts that were first assembled and distributed; thus it was the Colt Army Rifled Musket (as well as privately owned rifles) first used in battle by Berdan's sharpshooters. Notwithstanding, when the manufacture of the Sharps rifles was completed, the Colts were exchanged. Berdan, believing that the switch from the Colt to the Sharps may have angered Colonel Samuel Colt himself, wrote a correspondence to the Colts Arms Manufacturing Company, in care of its treasurer Hugh Harbison, extolling Colt's offering and explaining that he would not have made the switch had he not promised his men upon their enlistment that "they should have the Sharps rifles."[11]

The use of sharpshooters was not isolated to the North. Though the South did not have a "sharpshooters" designation, sixteen battalions were created and tasked with fielding long range riflemen.[12] Indeed, it was Jefferson Davis who, while serving as the American Secretary of War in 1853, advocated for the American Army's transition from smoothbore muskets to rifles and minie ball projectiles.[13] Nevertheless, the sharpshooters of the South were not nearly as distinguishable as their northern counterparts. If uniformed at all, the southern sharpshooters wore the same gray colors as their comrades, which proved to blend well with the natural environment. Despite Berdan's claims to the contrary, the southern sharpshooters surpassed their northern counterparts in terms of marksmanship abilities. This can more than likely be attributed to the men's backgrounds in hunting and tracking in the predominantly rural south.[14]

---

11. James L. Mitchell, *Colt: A Collection of Letters and Photographs about the Man, the Arms, the Company* (Stackpole, 1959), 264.

12. Pegler, *Sniper: A History of the US Marksman,* 80-81.

13. Antonio Scott Thompson and Christos G. Frentzos, *The Routledge Handbook of American Military and Diplomatic History: The Colonial Period to 1877* (London: Routledge, 2017), 323.

14. Ibid., 81.

Nevertheless, the South did not possess the human capital of the North and as such did not have the ability to field as many troops. Furthermore, the South did not have access to the same scale of manufacturing, namely weapons manufacturing facilities, and were required to acquire their arsenal by other means. While the small arms of the North were predominantly the Model 1861 Rifle Muskets manufactured by the Springfield Armory and Colt companies, to name two, the South was relegated to what they could beg, borrow, pick up off the battlefield, or import. That is not to say that the South did not manufacture weapons at all. There were a large number of independent and corporate armories that manufactured arms; however, the South paled in comparison to the industrial power of the North. In fact, some of the Springfield Model 1861 pattern rifles utilized by the Confederate Army were cloned and manufactured in the South. Of the imported rifles, one of British make was of particular interest and proved to be without equal insofar as accuracy was concerned. The Whitworth Rifle designed by Sir Joseph Whitworth featured hexagonal-polygonal rifling whereby the "twist of the Whitworth's bore imparted spin as the hexagonal bullet was fired."[15]   A test of the Whitworth in 1853 proved it accurate to 2,000 yards.

It is believed that a Whitworth rifle featuring a rudimentary telescopic sight was used by a Confederate sharpshooter to kill Union Major General John Sedgewick in the Battle of Spotsylvania Courthouse at a distance of over 800 yards in May of 1864. General Sedgewick is remembered in history for his infamous last words, "they couldn't hit an elephant at this distance."[16] General Sedgewick fell from his horse only a moment later, struck in the face by a Confederate bullet. Notwithstanding, the Whitworth Rifle was more expensive than any other single small arms weapon on the battlefield. Indeed, for the cost of one Whitworth Rifle, an entire company could be armed with Springfield clones. Before it gained notoriety, the Whitworth sold for approximately $100.00. Once its capabilities were understood, a single Whitworth Rifle could cost around $1,000.00. Adjusted for inflation, that is roughly $32,000.00 as of the penning of this text. It is no wonder that the Confederacy fielded relatively few and of those even fewer exist today. In October of 2017, a surviving Confederate Whitworth Rifle, still in its original packaging, sold at auction for $161,000.00.

---

15. Martin Pegler, *Sharpshooting Rifles of the American Civil War* (Oxford: Osprey Publishing, 2017), 14.

16. Martin McMahon, *Sedgwick Memorial Association* (Philadelphia, PA: Dunlap & Clarke Printers, 1887,) 78.

One Southern sharpshooter of note, a man by the name of John W. "Jack" Hinson, was far and away the deadliest "sniper" in all of the Civil War. At the outbreak of the Civil War, Hinson owned and lived on a plantation, known as "Bubbling Springs," in Stewart County, Tennessee. Hinson owned slaves but, for what it is worth, was against the South seceding from the Union. Nevertheless, he was described as a peaceable man who attempted to remain neutral, at least for the first two years of the war.

After Brigadier General Ulysses S. Grant took Forts Henry and Donelson in the Winter of 1862, he set up his headquarters at Hinson's immediate neighbor's, the Crisp Farm. The youngest Crisp, Hiram, recalled years later that Grant and his men were very kind to his family.[17] Hinson himself even hosted the General at Bubbling Springs on occasion. This all changed after Grant moved on. A Confederate commander, General Nathan Bedford Forrest, led men in a guerrilla campaign throughout the southern border states. These men became known as "bushwhackers," and were comprised of Confederate sympathizers and ad hoc militias. Under the command of Forrest, these men would materialize, ambush Union troops, and fade into the backcountry as quickly as they had appeared. As a result, Union officers and soldiers alike became highly suspicious of southerners which manifested as extreme paranoia, and in some cases . . . summary executions.

An example of the contempt with which bushwhackers were held can be found in a letter to Union Brevet Major General Clinton B. Fisk from legendary frontiersman Kit Carson, whose brother had been killed by bushwhackers. Carson sought to warn the general about inbound bushwhackers, stating of them "... these men have acted in such a way that good Union men as ever was are afraid of them... and I think they will do this country a great injury if they are permitted to scout."[18] General Fisk, who was headquartered in St. Joseph, Missouri, was, among his other duties, in charge of eliminating guerrilla factions in the south. He sent a telegram to Union Colonel J.T.K. Hayward regarding one of his subordinates, the commander of the Linn County, Missouri Militia, Captain Crandall. He ordered Crandall to "kill every bushwhacker he can put his hands upon, and to make the feeders, aiders and abettors of the villains sorry for what they have done to help on the iniquity..."[19]

---

17. Tom C. McKenney, *Jack Hinson's One-Man War* (Gretna, LA: Pelican Publishing Company, 2014), 77.

18. Robert N. Scott et al., *The War of the Rebellion: A Compilation of the Official Records of the Union and Confederate Armies.*, vol. 34 Part 4 (Washington: Government Printing Office, 1880), 316.

In Tennessee, while hunting one morning, Jack Hinson's sons George, 22, and John, 17, were mistaken for bushwhackers. The Lieutenant in charge of the Union patrol had the boys tied to a tree, shot, and beheaded. Their bodies were subsequently displayed in town and their heads placed on fence posts outside Jack Hinson's residence. Although Hinson had not considered the Union troops to be enemies up to this point, the murder of his sons and the depraved display of their bodies unleashed a whirlwind of revenge the likes of which no soldier, Union or Confederate, had ever seen. Hinson promptly commissioned the manufacturer of a custom rifle, its specifications being most similar to the legendary Whitworth Rifle. The custom rifle was crafted by William E. Goodman of Lewis County and featured a 41-inch barrel chambered in .50 caliber with a "Tiger Maple" stock.[20] In the Spring of 1863, Hinson put his new rifle to work, his first target . . . the lieutenant who had ordered his son's executions and beheadings. Hinson made a point of only targeting Union officers (occasionally making exceptions for bushwhackers.) Each time he killed a Union officer, he would stamp a small circle into his rifle, and by the time he completed his personal campaign against the Union, he had 36 small circles stamped on the weapon. This is not; however, an accurate number. The Union Army officially maintained that Jack Hinson was responsible for over 130 kills while some moderate estimates put it at just over 100.[21] Most of Hinson's targets were aboard river boats. From a perch atop a mountain hillside on the eastern side of the Tennessee River near the confluence of Hurricane Creek, Jack would engage Union soldiers on river boats trying to negotiate the Towhead Chute, a portion of the river that forced the riverboats to slow as they trudged against the river's current. This particular stretch of river would have required Jack to make between 100–500-yard shots at targets that were moving, albeit slowly. Nevertheless, Hinson killed Union officers with alacrity and completely controlled this portion of the river. On one occasion, he forced the surrender of an entire river boat, whose commander believed himself under attack from a Confederate Army unit, as opposed to one lone rifleman .

Despite the invaluable role they played in each of the armies, when the fighting ceased in 1865, what was left of the Civil War sharpshooters were "quickly mustered out of service and their rifles put into store. Their use, it was believed, had merely been a neces-

---

20. McKinney, *Jack Hinson*, 176.

21. Shahan Russell, "Jack Hinson: A Civil War Sniper Hell Bent on Revenge," August 11, 2019.

sary response to a peculiar set of circumstances..."[22] Much like the precision marksmen of the American Revolution, the sharpshooters of the American Civil War faded into history. Considering the psychological impact of being held in contempt by both friend and foe, many of the sharpshooters were content to fade. Berdan had satisfied his own appetite concerning notoriety and, citing medical conditions, had been discharged from his duties shortly after the events of Gettysburg. The 1st and 2nd USSS then fell under the command of Caspar Trepp who did not have the same ambitions as Berdan and viewed the sharpshooters as superfluous, relegating them to skirmishers and picket duties. Furthermore, Trepp held Berdan himself in contempt, going so far as to file formal complaints against Berdan for abandoning his men during Berdan's extended illness. Though later completely exonerated, this caused Berdan's exit to be mired in controversy. Trepp was soon thereafter killed in combat, his forces broken up and mustered into various Union units. Thus, the abolition of the 1st and 2nd USSS was not an officially sanctioned event; nor was it widely protested. Despite their invaluable contributions, the Sharpshooters of the North went out with a whimper. The Confederate Army was, of course, completely abolished. Jack Hinson, himself never having officially joined an army, returned to what remained of his family and moved to the village of White Oak Creek, Tennessee where he died in 1874. Due to the daunting tasks of reunification and reconstruction that lay ahead of the battered and bruised country, it is not surprising that none considered the next war. Nevertheless, the next war would come and would herald a new age in warfare. The traditional European warfare doctrine of the past would give way to new tactics dictated by new technologies that increased battlefield deaths by orders of magnitude. The birth cry of what could be considered the modern sniper would be the cacophony of buzzing machine gun fire and thunderous high explosive artillery barrage.

---

22. Pegler, *Sniper: A History of the US Marksman*, 89.

NOTES:

1. Martin Pegler, *Sniper: A History of the US Marksman* (London: Osprey Publishing, 2011), 71.

2. Pegler, *Sniper: A History of the US Marksman*, 72.

3. Ibid.

4. Hiram Berdan, "THE SHARPSHOOTERS OF THE NORTH," *The New York Herald*, May 31, 1861, Morning edition.

5. Pegler, *Sniper: A History of the US Marksman*, 77.

6. Ibid., 72.

7. Charles Augustus Stevens, *Berdan's United States Sharpshooters in the Army of the Potomac, 1861-1865,* reprint, (Bethesda, MD: University Publications of America, 1994), 10.

8. Caspar Trepp, "Letters from Captain Isler," *Papers*, accessed January 6, 2021.

9. Pegler, *Sniper: A History of the US Marksman,* 88.

10. "Col. Berdan's Sharpshooters," *New York Times*, August 7, 1861, p. 8.

11. James L. Mitchell, *Colt: A Collection of Letters and Photographs about the Man, the Arms, the Company* (Stackpole, 1959), 264.

12. Pegler, *Sniper: A History of the US Marksman,* 80-81.

13. Antonio Scott Thompson and Christos G. Frentzos, *The Routledge Handbook of American Military and Diplomatic History: The Colonial Period to 1877* (London: Routledge, 2017), 323.

14. Ibid., 81.

15. Martin Pegler, *Sharpshooting Rifles of the American Civil War* (Oxford: Osprey Publishing, 2017), 14.

16. Martin McMahon, *Sedgwick Memorial Association* (Philadelphia, PA: Dunlap & Clarke Printers, 1887,) 78.

17. Tom C. McKenney, *Jack Hinson's One-Man War* (Gretna, LA: Pelican Publishing Company, 2014), 77.

18. Robert N. Scott et al., *The War of the Rebellion: A Compilation of the Official Records of the Union and Confederate Armies.*, vol. 34 Part 4 (Washington: Government Printing Office, 1880), 316.

19. Ibid. 525.

20. McKinney, *Jack Hinson,* 176.

21. Shahan Russell, "Jack Hinson: A Civil War Sniper Hell Bent on Revenge," August 11, 2019.

22. Pegler, *Sniper: A History of the US Marksman,* 89.

PART II

Dawn of the Sniper

"You know how you smoke out a sniper? You send a guy out in the open, and you see if he gets shot. They thought that one up at West Point."

-Samuel Fuller, Film Director and WWII Veteran

CHAPTER III

WORLD WAR I

By the time the U.S. joined the fighting of the Great War in 1917, Europe had been engaged in warfare for nearly three years. Due to technological advances in artillery, namely the French 75 mm field-gun (nicknamed the "Devil Gun" by German troops); the introduction of Hiram Maxim's belt fed machine gun (used by both axis and ally alike); and Joseph Glidden's agricultural invention of barbed wire, new tactics were developed resulting in the implementation of trench warfare. The space between the trenches, known as "no-man's land," was a muddy hellscape. Battles would open with fierce artillery bombardments followed by men in the trenches going "over the top" of their trench and charging across no-man's land towards the enemy. This was met by raking machine gun fire from the enemy trench line. Those troops who would dive into craters to seek cover from the fire would soon find themselves trying to escape enemy gas attacks all the while navigating tangles of barbed wire. Considering the circumstances of past wars, having been comparatively mild to that which they were experiencing, it is no wonder that many believed that the Great War was to be the war to end all wars . . . the Apocalypse prophesied in the Revelation of St. John.

In order to fully understand the environment that the sniper mastered in World War I, one needs to understand what had changed between the time that the American Civil War ended, and World War I began. It seems that the collective American conscience places a mental chock after the Civil War, often with the presupposition that the Civil War more closely resembled the Revolutionary War than it did the first of the World Wars, especially insofar as tactics are concerned. That is not entirely inaccurate; however, it is not entirely accurate either. After all, the space of time separating the Revolution and the Civil War was 78 years, while the time between the Civil War and the first World War was only 49 years (53 years for the United States.) This is likely due to the mental images conjured when considering these wars. The Revolution, if described briefly, is characterized by its bright uniforms, heavy canons, horse mounted cavalry, and linear warfare doctrine. The American Civil War also offered non-combat conducive uniforms (especially by comparison to modern camouflage patterned uniforms), canons, horse mounted cavalry, and linear warfare doctrine. It would seem, indeed, that the Civil War has more in common with the Revolutionary War.

However, upon closer inspection, although the uniform colors had changed . . . the first of the World Wars still fielded horse drawn canons, mounted horse cavalries, and linear battle lines. In fact, it could be argued that Joseph Glidden's barbed wire may have saved as many lives as it took. Had no-man's land not been choked with it; it is plausible that commanders may have attempted to mount far more frontal assaults towards enemy lines than they did. Battles were still fought and won or lost by advancing on enemy lines in linear charges with the exception being that if the enemy lines were successfully taken, it was merely a waypoint to be occupied until the next charge. That is to say, neither a victor was declared, nor did the fighting stop. The victor of the battlefield was not necessarily predicated by a successful breakthrough of enemy lines.

So why does the first World War seem so different than every war preceding it? The answer is that the evolution of tactics is, and will always be, slower than technological development. After all, a tactic for negotiating a battlefield condition cannot be developed until the battlefield condition is known. The first World War brought with it never before seen technology that, by design, delivered death wholesale. Cheap and accurate metallic cartridge ammunition, machine guns, hydraulic recoil systems for canons, weaponized gas, barbed wire, the introduction of tanks, so on . . . The mental images of these alone make the Civil War and the first World War anything but comparable. Notwithstanding, the tactics themselves remained very similar. That is not to say that large phalanx formations would line up abreast and march in step towards enemy lines under a hail of withering fire, but much like the wars that took place in the eighteenth and nineteenth centuries, subsequent to a fierce artillery barrage, soldiers would charge en masse towards enemy lines. Thus, though the uniforms, weapons, and speed of attack changed, the tactics remained much the same. This failure to change tactics in the face of techno-logically superior war implements—at least in the early stages of the war—directly led to the stagnation of battlefield progress which, in turn, deteriorated into trench warfare. Although volumes could be written about the horrors experienced in the trenches, they proved to be second to none insofar as the incubation and honing of the sniper skillset.

Being confined to trenches in between attacks did not mean that the fighting ceased. The Great War would prove to be the birthplace of what could be considered the modern sniper. In one account, British Captain James Hyndson, commenting on the use of snipers by the Germans, wrote that a "gunner observing officer, who was with us all day directing his battery, is killed on his way out of the trench because he would not take sufficient precaution."[1] Indeed, the Kingdom of Germany was the first to readily embrace the skillset. In fact, the German troops began mounting steel plates along the top of their trench lines. The plates were placed at angles, creating a sawtooth pattern running the length of the German trench lines. The steel plates themselves would often have some fashion of sliding window whereby German *Scharfschüze* could fire upon British troops without exposing themselves to return fire.

This was the beginning of a method known as "loophole" sniping, an important skillset that modern snipers still work at mastering today. When the Great War began, the German Army recognized that "sniping would be a vital requirement in waging war"[2] and fielded them with alacrity. It was not long before the British responded in kind, creating their own brand of sharpshooters known as "snipers." The term "sniper" originated in British occupied India. A snipe was a swift game bird indigenous to India, and troops that were gifted marksmen capable of shooting a snipe became known as snipers. Still, "snipers" were merely individual soldiers who were gifted marksmen. It would take visionary leaders capable of recognizing a sniper's potential in warfare before snipers became what they are recognized to be today.

One such leader was British Major H. Hesketh Prichard, who obtained a commission at the outset of the war. Prichard, prior to the Great War, had been an avid large game hunter, professional cricketer, outdoorsman, and adventurer. Prichard recognized what the Germans were doing, as well as the need for establishing professional British snipers.

1. James G. W. Hyndson, *From Mons to the First Battle of Ypres* (United States: Pickle Partners Publishing, reprint 2015), 94.

2. Martin Pegler, *Out of Nowhere: A History of the Military Sniper, from the Sharpshooter to Afghanistan* (Oxford: Osprey, 2011), 83.

Furthermore, he recognized that snipers could be utilized as force multipliers. He, of course, wanted to create gifted marksmen, but also a class of soldier who could move stealthily, blend in with the environment, and gather intelligence.

Prichard, much like Berdan half a century before, was enthusiastic about establishing and fielding snipers. However, also like Berdan, Prichard's enthusiasm was not shared by all. Nevertheless, with this concept in mind, Prichard established the first formal sniper training program in history. This was not an easily accomplished task as "many senior officers . . . refused to accept the usefulness of employing snipers, regarding them as un-mitigated nuisances."[3] Just another of many examples that exist throughout the history of snipers demonstrating the regard with which they were held. Despite the difficulty, Prichard was successful at establishing the First Army School of Scouting, Observation and Sniping. In addition to marksmanship skills, Prichard emphasized intelligence gathering and advanced camouflage concepts. Insofar as camouflage is concerned, he was quite fond of the techniques used by the Lovat Scouts.

The Lovat Scouts were comprised of Scottish Highland ghillies (or gillies.) Traditionally, a ghillie was an outdoorsman servant who accompanied a Scottish lord for a number of reasons, to include being a stalker on hunting excursions. In order to stalk the prey, the ghillies would dress in animal skins, or in some cases, garb to which they would attach natural occurring foliage. In 1900, during the course of the Second Boer War in South Africa, the Lovat Scouts were established by the 14th Lord Lovat, Major-General Simon Joseph Fraser, who chose an American, Frederick Russell Burnham, to lead the unit. The Lovat Scouts were noted for being the first military unit to utilize the ghillie suit. Prichard believed that "no man, unless very skilled in observation, could spot a hidden Lovat scout wearing one from a distance of 10 yards."[4] It is not surprising that he would rely on the Lovat Scouts, now headed by the 15th Lord Lovat, Major Simon C. J. Fraser, to train his snipers.

He recalled in his text *Sniping in France* that the Lovat Scouts were extremely gifted in observation and that during the course of training, his students were asked to identify a group of troops some 600 yards away. The students reported that they observed some troops in blue uniforms; however, one Corporal Cameron, a Lovat Scout, reported that

---

3. Martin Pegler, *Sniping in the Great War* (Barnsley: Pen & Sword Military, 2017), 23.

4. Pegler, *Out of Nowhere,* 135.

they must be Portuguese troops. He reasoned that, "They must be either Portuguese or French... and as they are wearing the British steel helmet, they must be Portuguese."[5]

In 1917, citing the Zimmerman Telegram and the sinking of the RMS Lusitania, America became involved in the war. What had been a European Great War, now became a World War. Upon arrival in Europe, Americans immediately became involved in trench warfare, and by extension, sniper warfare. Prichard later commented that the American snipers "were also fine shots, and thoroughly enjoyed their work..."[6] This sentiment was echoed by Marshal of France General Philippe Pétain who commented on the "partiality of the American soldiers for sniping (in which they easily excel) ..."[7]

American snipers learned how to design ghillie suits, perform stalks, gather intelligence, and engage enemy combatants at a distance. The psychological effect of employing a sniper was devastating to an enemy. Thus, much like the American Civil War, snipers who were caught by enemy troops were summarily executed on the spot by both sides. Understandably, sharpshooters/snipers are so feared and despised, that this practice has been continuous since the American Revolution. Nevertheless, the U.S. seemingly bought in completely to the concept of facilitating snipers. The Springfield M1903 bolt-action rifle that was standard issue to soldiers began a process of upfitting.

New advances in optics saw telescopic sights being retrofitted to M1903s which were then repurposed as sniper rifles. In an age of warfare that seemed to be characterized by machine gun fire and artillery bombardments, an extensive amount of capital was being invested in making the sniper better. One American sniper, an army Private named Herman Davis, was thirty years old when the United States entered the war. Private Davis, who had been born and raised in Manila, Arkansas, a rural area approximately twenty miles west of the Mississippi River, had spent a significant portion of his life hunting and shooting in the forests and around the banks of the Mississippi. Davis had been drafted into the Army in 1918 and was assigned to I Company, 113th Infantry Regiment, 29th Division of the American Expeditionary Force.

While serving in France, Davis earned the Distinguished Service Cross for assaulting a German machine gun nest near Mollville Farm during the Meuse-Argonne Offensive.

5. H. Hesketh-Prichard, *Sniping in France* (Barnsley: Pen & Sword Military, reprint 2014), 75.

6. Hesketh-Prichard, *Sniping in France*, 90.

7. Phillipe Petain, "Training of American Units with French," *United States Army in the World War, 1917-1919* (University of California, Berkeley, 1989), 295.

On another occasion, Davis observed a German machine gun nest being set up by five German soldiers. Davis was told that the nest was ranged at 1,000 yards and was beyond the range of small arms fire. Armed with an M1903 Springfield Rifle with open sights, Davis replied, "Why, that's just a good shooting distance."[8] He then proceeded to engage and eliminate all five German soldiers occupying the nest.

Indeed, American troops, particularly those who had lived in the American south, proved to be adept snipers. Perhaps the most notable of these was Sergeant Alvin York, who had been raised in rural Tennessee. York grew up hunting in the mountains of Appalachia and, like Private Davis, had been drafted into the Army upon the United States' entry into what would become World War I. Though York struggled with the concept of fighting, due in large part to his Christian faith, York proved to be an exceptional soldier. He had been assigned to G Company, 2nd Battalion, 328th Infantry of the 82nd Division of the American Expeditionary Force and, like Davis, took part in the Meuse-Argonne Offensive. On October 8, 1918, Corporal York, with his unit comprised of sixteen other men, was ordered to capture Decauville Railroad in the Argonne Forest. Unfortunately the map they had been provided was entirely in French, a language that none of the soldiers spoke. As a result, York's unit quickly found themselves behind enemy lines. After winning a small engagement with a German unit, the surviving German soldiers began calling for help. More than half of York's unit, to include all the non-commissioned officers, were immediately cut down by two German machine gun nests. York took command of the six remaining soldiers in his unit and ordered them to begin firing on the nests. Meanwhile, York himself began assaulting towards the nests . . . alone.

Being a gifted marksman, York set to work picking off the soldiers manning the machine gun nests as he worked his way towards them. When the German soldiers realized what was happening, a fire team of five or six German soldiers charged towards York, who drew his pistol and eliminated each in turn. He then successfully reached the machine gun nests, eliminating each crew. The German commander subsequently surrendered to York, who . . . along with his six remaining men . . . took 132 German prisoners and returned to American lines.

York was immediately promoted to Sergeant and subsequently received the Medal of Honor, French Croix de Guerre and French Legion of Honor. Sergeant York, who had

---

8. Martin Pegler, "The Allies Strike Back: The Genesis of Sniping, Part 5," An Official Journal Of The NRA, May 25, 2017.

been in Europe less than five months, returned home a hero and is recognized as one of the most decorated soldiers of World War I. Although Sergeant York was not formally trained to be a sniper, the details of Sergeant York's assignment to capture the railroad contain all of the elements of a sniper's job; a small unit tasked with scouting and taking a salient position by eliminating the enemy. Notwithstanding the fact that Sergeant York's unit did not accomplish the initial task, his ability to remain calm in the face of dire circumstances, provide accurate return fire, and subsequently defeat an enemy with an *extreme* numerical advantage has all the hallmarks of a sniper. That is to say, Sergeant York and his men became, by necessity, force multipliers. Thus, while he may not be formally recognized as a sniper, he most definitely fit comfortably among their ranks.

It would be a great disservice to discuss American snipers who were involved in World War I without giving due regard to the weapon system they utilized. The Springfield Model 1903 was a five-round internal magazine fed bolt action rifle chambered in the vaunted .30 Caliber 1906 (.30/06) cartridge. The lay person, when asked to name an iconic American military weapon, may first name Eugene Stoner's Armalite Rifle 15 (AR-15) later designated the M16 . . . or perhaps John Garand's offering the Garand M1; however, a sniper will almost always list the 1903 Springfield. It must first be understood that a professional sniper will approach a precision weapon system in the same way that a sommelier will approach a Cheval Blanc, or an art collector approaches a Monet. Every detail and line is appreciated. However, unlike a fine wine or painting, the simplicity of the 1903 is what may be most beautiful . . . at least to some.

Many snipers, if not all, are simplistic in the belief that form follows function. In this respect, the M1903 is a work of art . . . truly a case study in getting it right the first time. It should be noted that the M1903 is so beautiful and iconic that it is still utilized ceremonially by honor guards today. Notwithstanding, beyond the aesthetic appeal of the rifle, what mattered most to snipers in World War I was its accuracy. The design of, and subsequent switch to, the M1903 was a response to the shortcomings of the Krag-Jørgensen rifle utilized by Americans during the Spanish-American War in 1898. Although the Krag was an accurate rifle, the internal magazine was loaded one round at a time and featured .30/40 Krag ammunition. The rifle was outmatched by the Spanish Mauser in every way. Once the M1903 was adopted and became the standard issue rifle to U.S. servicemen from 1903 until 1936, it was snipers who proved able to coax maximum efficiency from the rifle.

When the M24 SWS and the M40, both featuring a Remington 700 action, became the standard issue to snipers within the American military in the late twentieth century, the featured .308 Winchester cartridge reduced the maximum effective range to 800 meters (although the cartridge itself, in truth, is capable of going beyond 1,000 meters.) In contrast, the M1903, featuring a modest velocity of 2,800 fps, had an effective firing range of 1,100 meters on a point target. Thus, the rifle . . . without considering any telescopic optics . . . was one size fits all, ideal for any logistician. That is not to say that the M1903 is better suited to the task than are modern rifles. One of the shortcomings of the M1903 that was less well understood then was its wood furniture.

Although wood handguards surrounding the barrel made the rifle sturdier and heavier—capable of mitigating more recoil—it also eliminated the concept of a free-floated barrel, a characteristic that is absolutely fundamental for modern precision rifles. Aside from the rifle, telescopic optics were seeing unprecedented use. At this point, optic technology was nowhere near modern standards, but neither were they new technology. Telescopes had been retrofitted to rifles since the mid-nineteenth century, and it is speculated that the Confederate sharpshooter who engaged and killed General John Sedgewick at the Battle of Spotsylvania Courthouse during the American Civil War had been utilizing a telescope affixed to his rifle. Indeed, the engineering of telescopes outpaced advancements in firearms, meaning that though telescopes existed, it was not until the mid-nineteenth century that mass produced rifles proved accurate enough to utilize them. However, telescopic optics up to, and for some time after, this point were incredibly delicate, somewhat unreliable, and extremely cantankerous accoutrements.

In 1913 the Warner & Swasey Company of Cleveland, Ohio updated their Model 1908 and created the Warner & Swasey Co. Telescopic Musket Sight Model of 1913 that attached to the M1903 rifle. The prismatic scope was an upgrade to their Model 1908 offering and was nearly identical to the Model 1908. It was solid brass construction, featured a rubber eye cup and, what is perhaps its greatest flaw, was offset to the left of the upper receiver to allow loading the rifle via stripper clip. This made the rifle somewhat difficult to fire as the sniper was not able to comfortably rest his head on the comb of the stock when firing. While the Warner & Swasey Model 1913 was the standard issue to soldiers in the U.S. Army, the U.S. Marine Corps opted to utilize the Winchester A5 model telescopes that more resembled traditionally mounted rifle scopes. The Winchester scopes had not been procured for sniper use; rather they were to be utilized by the Marine Corps' competitive shooting team.

As marines deployed to Europe; however, these competition M1903 rifles with A5 scopes were made available to them. The Winchester was of a sturdier construction compared to the Warner & Swasey 1913 and as previously mentioned, were mounted vertically centered above the receiver. Both optics were engineered for five-power magnification and proved well suited to the task of sniping, providing a relatively large field of view at range. By wars end, the Warner & Swasey 1913 had been mounted to roughly 1,500 combat operational rifles with the Winchester A5 being mounted on roughly 500. Though many advances would take place in optics, the M1903 proved accurate enough to satisfy the role of sniper rifle throughout World War I and would go on to see further use in World War II and the Korean War.

When the war ended, on the 11th hour of the 11th day of the 11th month in 1918, American snipers were sent home, their training programs abandoned, and their tactics perceived as suitable only for trench warfare. Much like the Civil War, U.S. snipers were believed to be a response to a peculiar and isolated set of circumstances. To say that American military leadership were unaware of the success of American snipers would be highly disingenuous considering the amount of capital that had been invested in them. Today's Army Sniper Association writes that, "As a result of the First World War sniping became a recognized part of the military machine."[9] While that may be true, the recognition that snipers at that time received could not be characterized as "good."

The ASA provides insight, albeit unknowingly, as to how snipers were perceived . . . both then, and to some degree, now. The organization states, "trained marksmen would function essentially as assassins, often targeting any moving object behind enemy lines, even if they were engaged in peaceable tasks (which meant that if a sniper was taken prisoner, he could expect no mercy on either side.)"[10] The characterization of "assassin" is, perhaps, the salient word of that particular statement. Soldiering is considered an honorable profession; however, an assassin is collectively perceived as an anathema.

Thus, it is arguable that the contempt with which the profession was held—and by extension the men who were part of it—resulted in its demise. After all, no professional western military wants to be associated with participation in nefarious deeds. Although their units were disbanded, their weapons went into stores, indicating leadership's quiet acknowledgement that their use might be needed once more.

---

9. "History: Army Sniper Association," Army Sniper Association | Army Sniper Association, June 2, 2017.

10. Ibid.

NOTES:

1. James G. W. Hyndson, *From Mons to the First Battle of Ypres* (United States: Pickle Partners Publishing, reprint 2015), 94.

2. Martin Pegler, *Out of Nowhere: A History of the Military Sniper, from the Sharp-shooter to Afghanistan* (Oxford: Osprey, 2011), 83.

3. Martin Pegler, *Sniping in the Great War* (Barnsley: Pen & Sword Military, 2017), 23.

4. Pegler, *Out of Nowhere,* 135.

5. H. Hesketh-Prichard, *Sniping in France* (Barnsley: Pen & Sword Military, reprint 2014), 75.

6. Ibid., 90.

7. Phillipe Petain, "Training of American Units with French," *United States Army in the World War, 1917-1919* (University of California, Berkeley, 1989), 295.

8. Martin Pegler, "The Allies Strike Back: The Genesis of Sniping, Part 5," An Official Journal Of The NRA, May 25, 2017.

9. "History: Army Sniper Association," Army Sniper Association | Army Sniper Association, June 2, 2017.

10. Ibid.

CHAPTER IV

WORLD WAR II

World War II is perhaps the most studied conflict in human history. Within the United States one need only access a video streaming application and a brief search will produce scores of results . . . fiction and non-fiction . . . concerning anything and everything having to do with World War II. Network television companies have produced thrilling programs concerning everything from secret government plots to introduce estrogen into Hitler's diet, an attempt to make his voice higher in pitch and his iconic mustache fall out—it was never carried out—all the way to three-dimensional computer animated recreations of the Battle of Midway, the turning point of the war in the Pacific. Many journalists, professors, television personalities, veterans, and amateur historians have written countless books concerning everything from the entirety of the war to firsthand accounts of individual battles.

Indeed, the American appetite for consuming knowledge of World War II is voracious. Still, finding source material concerning American snipers in World War II, or any war prior to Vietnam, is a daunting task. It might be likened to analyzing a crime scene, in which a void is observed in a blood spatter pattern, indicating that something or someone had occupied a particular space at the time of the event, but who or what remains to be discovered. Often, it is only found in the occasional autobiography, or the recollection of a soldier being interviewed decades after the events. This is puzzling as, for the most part, Americans have always celebrated their soldiers.

The American experience for most individuals, with the exception of a few counter-culture movements, has been collective pride in the American soldier, especially for the last two decades. Regardless of political affiliation, race, religion, gender, or social status, Americans, by and large, agree that saying thank you to a soldier, sailor, marine, airman, or veteran is the right thing to do. It provides a sense of altruism. However, most Americans do not understand the complex nature of warfare, or the emotions navigated by those who volunteer to do violence on the behalf of others. It is arguable that most do not want to understand. That is not meant to be demeaning, or to imply that there is a lack of empathy among the population at large. It is merely an observation of the many differences that accompany the human condition. In short, most want to know that the job is getting done; the details about *how* are of little to no importance and only matter if the event was reprehensible, such as the genocide of a village or other atrocities.

It may seem as though none of this has anything to do with American snipers in World War II. Indeed, it could be considered a pessimistic rant. That is not the intention although it may have been the result. Consider the following—one may notice that the name Vasily Zaitzev is somewhat familiar—especially to cinephiles and war historians at the very least. He was a Soviet sniper during World War II and the focus of the 2001 feature film *Enemy at the Gates* starring Jude Law, Rachel Weisz, and Ed Harris. Zaitsev was the recipient of the Hero of the Soviet Union medal and was widely celebrated within the Soviet Union then, and in Russia now. So harrowing were his exploits during the Battle of Stalingrad that an American film was made celebrating his feats nearly six decades later. However, even the American war historian, amateur or otherwise, would be hard pressed to name a single American sniper who participated in World War II. The argument might be made that Zaitsev's story, recorded in his autobiography, was compelling. Thus it is fitting that Hollywood would produce a dramatic tale detailing his exploits although there was no shortage of creative license utilized. Maybe so, but the question remains, why was Zatisev comfortable publishing his exploits while American snipers of that era were not? Herein lies the void in the pattern, whereby their existence is a certainty; however, their deeds are all but lost.

It is safe to say that the very first step the world took towards World War II was the signing of the peace treaty that ended the first world war. By the time the war erupted in 1939, Adolf Hitler had invested a considerable amount of capital in the German war machine, to include building a large portion of his tanks inside Russian borders. This was accomplished via a non-aggression pact signed by Hitler and Stalin. Despite this agreement, Hitler launched Operation Barbarossa on June 22, 1941, invading Russia utilizing the same *blitzkrieg* tactics that had so quickly toppled Western Europe. Although the *Wehrmacht* initially found victory, in August of 1942, Hitler's powerful Sixth Army was halted at an industrial city located on the Volga River. The city, Stalingrad, was named in honor of the Soviet Premier Joseph Stalin, and arguments still abound as to whether Hitler's justification for trying to take it were tactically motivated or more oriented towards a propaganda victory. Regardless, what Hitler's Sixth Army found in the rubble-choked urban labyrinth was a new brand of warfare that the German's called *Rattenkrieg,* or War of the Rats. The Soviets, on the other hand, found an incubation chamber of sorts, an environment conducive to developing the most formidable snipers the world had ever seen.

Furthermore, the Soviet Union did not discriminate insofar as who was allowed to pick up a rifle and fight. Men and women alike rushed to fight for "Mother Russia," many of them finding their way into sniper schools. At Stalingrad, legends like Vassili Zaitsev and Lyudmila Pavlychenko thrust their way into the history books. In stark contrast to American snipers, Zaitsev and Pavlychenko, who both had staggering kill counts, were heralded as heroes of the Soviet Union. While plenty has been written on Zaitsev and Pavlychenko, and plenty more could be, this text is meant to focus on the American sniper experience. Though, exploring the Soviet sniper's postwar experience and how it contrasts with that of the American sniper's plight casts a spotlight on the shortcomings of the U.S. military and the American public, at least insofar as the sniper is concerned.

In the United States, soldiers and marines were preparing to ship to England and the islands of the Pacific. The naval base at Pearl Harbor in Honolulu, Hawaii had been attacked by Japanese fighter and torpedo planes on Sunday, December 7, 1941. In response to this attack, the United States declared war on Japan on December 8, 1941. Three days later, Nazi Germany declared war on the United States and the U.S. Congress reciprocated. The primary weapon issued to the troops was the M1-Garand Rifle designed by John Garand. The M1 was a semi-automatic, internal magazine-fed rifle chambered in .30-06. When the rifle would fire its last round, it made a "ping" sound. This was caused by the ejection of the en bloc clip, a thin piece of metal that held the cartridges together. Both the "clip" and rounds were loaded into the internal magazine as a single unit. When the last round was fired, the clip would be ejected, striking the metal receiver, creating the characteristic audible "ping."

Men began training, and although the basic training environment that existed was far from what it would become, the fighting men of the United States were prepared for war. Nevertheless, programs associated with training, equipping, and fielding snipers did not exist. Leadership believed it would be too costly an endeavor.[1] This sentiment was seemingly shared by the U.S.'s British allies and may have been best characterized by Lieutenant Colonel Nevill A. D. Armstrong, a British officer and sniper school instructor during World War II, when he commented of the British military's regard for the discipline that, "There appeared to be a tendency amongst Army musketry men to scorn the sniper—they held that sniping was only a 'phenomenon' of trench warfare and would be unlikely to occur again."[2] Nevertheless, on the battlefield, enemy snipers were proving that the need for advanced marksman within the U.S. military was necessary. Instead of investing in the development of a designated sniper weapon system, the ordnance department was determined to use what already existed . . . the M1 Garand.

This was a difficult endeavor, as the M1 was loaded from the top of the receiver, precluding the use of a top-mounted scope. To solve the problem, the war department mounted the scope precariously to the left side of the receiver of the rifle. The scope, a commercial production Lyman Alaskan, was redesignated the M84 scope. The sniper version of the Garand was designated the M1C. At first look, it appeared as though the problem had been solved; however, aside from being unwieldy, at long range, mounting a scope anywhere other than directly over the center of the bore line presents mathematical problems with angular measurement. Establishing a firing solution for a long-distance shot with an offset scope produced a parallax error that could not be corrected via the technology available at the time. Furthermore, the characteristic ping of the clip being ejected could give a sniper's position away or alert the enemy that he had fired his last round. These problems resulted in most snipers preferring the Springfield M1903 from World War I. The old Springfield rifles were pulled from storage, retrofitted with modifications, and put into service once more. The redesignated M1903A4's proved accurate, precise, and reliable.

While one problem had been solved, insofar as supplying the necessary equipment, there were still no formal training programs to speak of. Often, those who were considered

---

1. Pegler, *Sniper: A History of the US Marksman*, 129.

2. Nevill A. D. Armstrong, *FIELDCRAFT, SNIPING AND INTELLIGENCE*, Reprint (E. Sussex: Naval & Military Press, 2019).

the best shot in a unit were assigned to be snipers. In a correspondence, Corporal J. D. Penney of the 82nd Airborne, who had served in North Africa and Western Europe during World War II, discussed the methods he utilized to teach himself how to use his issued sniper rifle. He stated, "It took weeks for us to become proficient, and we ended up being pretty good, but we never got training."[3] This was the case throughout the U.S. military insofar as snipers were concerned, the only exception being individual commanders who took it upon themselves to provide training. Furthermore, the condemnation that U.S. snipers/sharpshooters had faced since the American Revolution was still manifest within the ranks. By extension, captured snipers rarely became prisoners of war. As Pegler points out, "by the time the fighting moved into France, there was no longer tolerance for Nazi snipers, and no quarter was asked or given. Captured snipers, of either side, would be routinely shot..."[4] This, having been the case throughout the history of warfare involving firearms, seemed to be widely known . . . and accepted as a condition of being a sniper. Still, it seems peculiar that this comes with the territory given that a sniper discriminately selects individual targets as opposed to indiscriminately delivering raking machine gun fire at silhouettes in the distance or delivering a highly explosive incendiary device from 30,000 feet . . . both of which kill en masse. Penney reflected on fellow soldiers and the regard with which they held their own snipers, stating:

> Many of our buddies didn't like what we did and called us 'ten cent killers' [a reference to the cost of a cartridge] but when they were pinned down by a Kraut sniper in the hedgerows it was always 'Get a damn sniper up, quick' and they were happy enough that we dealt with them. Then we were hero's [sic] for ten minutes.[5]

Ernie Pyle, an American war correspondent and journalist who documented World War II, once wrote, "Sniping, as far as I know, is recognized as a legitimate means or warfare. And yet there is something sneaking about it that outrages the American sense

3. J.D. Penney, Correspondence with the Author, *Sniper: A History of the US Marksman*, October 2011, 131.

4. Pegler, *Sniper: A History of the US Marksman*, 133.

5. Penney, Correspondence with the Author, *Sniper: A History of the US Marksman*, 134.

of fairness."[6] It is worth noting that Pyle, when writing this observation, did not differentiate between soldier and civilian. It is fair to assume that, being a journalist, he was careful and concise with his language. Despite being surrounded by soldiers . . . of his own nation as well as others . . . Pyle noted that "Americans" took exception. Considering the context of his work, he very well could have been writing of American soldiers specifically, utilizing the "American" identifier as a generalization to differentiate between American soldiers and their British allies and German enemies in the European theater. On the other hand, Pyle often adds the qualifier "soldier" throughout the text when speaking specifically about American troops. In any event, he was universally loved by American readers back home in the U.S. and had a large readership. An argument could be made that he was attempting to innocuously educate Americans to the realities of war and of the invaluable contribution of the sniper . . . but most likely this is not the case. He would later write in the same text, "The average American soldier had little feeling against the average German soldier who fought an open fight and lost. But his feelings about the sneaking snipers can't very well be put into print."[7] Adding the qualifier "sneaking" seems to shed light on Pyle's own feelings towards the discipline. Unfortunately, the answer will remain in the realm of speculation . . . and the lesson may never be imparted as Pyle, himself, was later targeted and killed by a German sniper.

Meanwhile, in the Pacific theater, the U.S. Navy and the U.S. Marine Corps were involved in an island-hopping campaign championed by General Douglas MacArthur. While the United States Army was in control of the European theater, this was the Marines' territory. Notwithstanding, the Marines were facing similar problems. While the Army struggled through the hedgerows of the Bocage in France, the Marines were staring down the dense jungle vegetation that was common on each and every Pacific Island encountered. Unlike the Army, the Marines were facing an altogether different enemy. What Imperial Japan lacked in mechanized warfare implements, they more than made up for in warrior spirit. In the Japanese, the Marines encountered an enemy that did not seem to have any concept of surrender. Indeed, the Japanese soldier would fight to the very last man. Unlike the Nazi in Europe, the Japanese were not entrenched in an ideology. It was more than that. It was part of their culture, stretching back centuries, dictating that it was better to die honorably than to live with the shame of defeat. This made the

---

6. Ernie Pyle, *Brave Men*, Reprint (Lincoln, NE: University of Nebraska Press, 2001), 397.

7. Pyle, *Brave Men*, 398.

Japanese sniper an especially dangerous foe, whereas a German sniper would kill "as many Americans as they could and when their food and water ran out, they surrendered..."[8] the Japanese sniper would not surrender, but instead, draw a sword and charge headlong towards enemy troops in a final act of honorable sacrifice. To compound this, the fighting that took place on the islands of the Pacific was intimate, close quarters, jungle warfare.

Among the Marines fighting on these islands, was Colonel William "Wild Bill" Whaling, who had followed a non-traditional path to his commission. He had enlisted during World War I and served with distinction in the American Expeditionary Force at Thiaucourt, France between September 12 – 26, 1918. Whaling earned a commission to Lieutenant . . . and later received a Silver Star Citation for gallantry on his World War I Victory Medal.[9] By the outbreak of World War II, Whaling had been promoted to Colonel and was tasked with leading a specialized unit at Guadalcanal, of the Solomon Island chain. As is often the case, Whaling was a competitive shooter long before the events of World War II would transpire. Furthermore, he had gained expertise in jungle warfare. Whaling was tasked with training and leading a Scout Sniper section that would be the "tip of the spear" during an operation to push towards the Japanese controlled Matinakau River in the north of Guadalcanal.

Unlike many commanding officers, past and present, Whaling's leadership of this specialized unit was quite literal and unique. Whereas some commanders would have facilitated the training of the unit and subsequently dispatched them with orders, on November 1, 1942, Whaling set out with his men into the thick jungle, leading from the front. Whether this was admirable or foolish can be speculated on by others. Senior officers may argue that a commander's role is logistical and administrative in nature and should focus on generating tactical plans and responses in the rear. At the same time, the average soldier may find inspiration in such a deed.

Regardless of whatever conclusion might be drawn, Whaling's force, known as Whaling Group, saw success. For his part, Whaling received the Legion of Merit, the citation reading: "Colonel Whaling organized a scout-sniper detachment and supervised the training of selected groups in scouting stalking and ambush tactics."[10] Considering that Whaling set the groundwork for what would become the U.S. Marine Corps' elite

---

8. Pyle, *Brave Men*, 398.

9. "Whaling, William John 'Wild Bill,'" tracesofwar.com, accessed August 1, 2022.

10. "William Whaling - Recipient," The Hall of Valor Project, accessed August 1, 2022.

fighting units, namely their Recon and Scout/Sniper Programs, it is strange that Whaling Group did not receive more attention, then or since. This could be due to a number of factors. First, Whaling Group was providing flank security for the 1st Battalion, 7th Marine Division, which was being led by Lieutenant Colonel Lewis B. "Chesty" Puller. He would retire in 1955 as the most decorated Marine in U.S. history . . . making the shadow he cast broad and far reaching. Furthermore, Puller's unit encountered stiff resistance during the Matinakau offensive, whereas the resistance Whaling's men encountered was comparatively light. Second, the jungle fighting in which Whaling was involved was the antithesis of what most would consider the hallmarks of sniper warfare, namely long-range engagements, though it was absolutely the stock and trade of recon. Third, but not least, the prevailing disposition felt towards snipers. The reality may be that it does not boil down to one reason or the other, but instead is a combination of all three and more. An existentialist would reason that if something exists, then it inevitably has an impact on its surroundings, great or small. Perhaps that is the case. Nevertheless, if any of this bothered Whaling, he did not let it show. He would continue to serve in the military, eventually as a Brigadier General during the Korean War.

With the surrender of Nazi Germany, all that remained for bringing about the end of the second World War was defeating Japan. Doing so presented an extremely difficult ethical dilemma. President Franklin D. Roosevelt had passed away prior to the end of the war, and President Harry S. Truman was now holding the reigns. He was almost immediately faced with a difficult choice—knowingly send a million American troops to their death, or deliver devastation, the likes of which had never been seen, to enemy combatant and civilian alike. The choice he made is well known.

The mention of Hiroshima and Nagasaki immediately generates images of giant mushroom clouds rising into the sky above Japan. It is impossible to accurately tabulate the exact number of casualties generated by the two explosions. A number of factors are involved, including deaths caused by radiation fallout (long or short term). However, for the purpose of this text, focus will be placed on only those who were immediately killed by the blast. The Manhattan Engineer District's estimate, as outlined in the Oughterson Commission study, states of the blast in Hiroshima alone: "The death toll of the first day will never be accurately known. Perhaps it was between 40,000-50,000."[11] Nagasaki,

---

11. A. W. Oughterson et al., "Medical Effects Of Atomic Bombs The Report Of The Joint Commission For The Investigation Of The Effects Of The Atomic Bomb In Japan," 1 Medical Effects Of Atomic Bombs The Report Of The Joint Commission For The Investigation Of The Effects Of The Atomic Bomb In Japan § (1951).

although having a smaller population, was comparatively similar, with some 20,000 immediately killed by the blast.[12] If only those killed by the immediate detonation of both bombs is tabulated, that is somewhere between 60,000 to 70,000 lives extinguished in the space of a heartbeat. Still, the photo of Colonel Paul Tibbets waving from the cockpit of the first aircraft to drop an atomic weapon, an aircraft named after his mother, Enola Gay, is iconic. Far from scorn, it is regarded with a degree of pride and, perhaps sorrow. Tibbets himself was heralded as a hero. After all, the alternative was an amphibious assault of the Japanese mainland, the casualties of which were estimated at one million, and that was just Americans. Ethically speaking, Truman took a course of action that potentially saved more lives than it took. Ethics aside, war is ugly. Notwithstanding, the bombings of Hiroshima and Nagasaki sealed the fate of Japan. Emperor Hirohito formally surrendered six days after the bombing of Nagasaki, on August 15, 1945.

However, as predicted by the Napoleonic philosopher and soldier Carl von Clausewitz, peace is merely the space between wars and the U.S. would face another war in less than a decade. Tensions developing between the Soviet Union and the United States would result in the stalemate of the Cold War. Proxy wars would erupt in Asia and, in the name of avoiding a nuclear holocaust, the U.S. would find itself abandoning total war doctrine for a "limited war" approach. Although sniper programs continued to struggle to gain traction, the future of warfare would present an environment uniquely conducive to the sniper's trade. Unfortunately, all the lessons learned would need to be re-learned as "Snipers had been needed, they were used, and now it was better not to talk about such cold killers in the rational light of peace."[13]

---

12. Ibid.

13. Charles W. Sasser and Craig Roberts, *One Shot, One Kill* (New York: Pocket Books, 1990), 83.

NOTES:

1. Pegler, *Sniper: A History of the US Marksman,* 129.

2. Nevill A. D. Armstrong, *FIELDCRAFT, SNIPING AND INTELLIGENCE,* Reprint (E. Sussex: Naval & Military Press, 2019).

3. J.D. Penney, Correspondence with the Author, *Sniper: A History of the US Marksman*, October 2011, 131.

4. Pegler, *Sniper: A History of the US Marksman,* 133.

5. Penney, Correspondence with the Author, *Sniper: A History of the US Marksman*, 134.

6. Ernie Pyle, *Brave Men*, Reprint (Lincoln, NE: University of Nebraska Press, 2001), 397.

7. Ibid., 398

8. Ibid.

9. "Whaling, William John 'Wild Bill,'" tracesofwar.com, accessed August 1, 2022.

10. "William Whaling - Recipient," The Hall of Valor Project, accessed August 1, 2022.

11. A. W. Oughterson et al., "Medical Effects Of Atomic Bombs The Report Of The Joint Commission For The Investigation Of The Effects Of The Atomic

Bomb In Japan," 1 Medical Effects Of Atomic Bombs The Report Of The Joint Commission For The Investigation Of The Effects Of The Atomic Bomb In Japan § (1951).

12. Ibid.

13. Charles W. Sasser and Craig Roberts, *One Shot, One Kill* (New York: Pocket Books, 1990), 83.

PART III

The Cold War

"God created war so that Americans would learn geography."

- Mark Twain

CHAPTER V

THE KOREAN WAR

Subsequent to the end of World War II, the United States and the Soviet Union gazed suspiciously towards one another. The iron curtain fell across eastern Europe, and Russia, covetous of the United States' development of an atomic weapon, began their own nuclear programs. In the meantime, the proliferation of communism throughout the far east became a major concern for the United States. What would follow was a series of limited wars throughout the Far and Middle East, pitting the U.S. against the Soviets by proxy, the first of which erupted in Korea, a small peninsula that separates the Yellow Sea and the Sea of Japan.

In the colossal shoving match that made up the first year of the Korean War, the North, prior to the United States' entry into the battle, had nearly wiped out the South. Under the leadership of General Douglas MacArthur, the United States Army joined the fray on June 27, 1950 and after an amphibious assault of Incheon in September of that year, and no small amount of the iron resolve displayed by General Walton H. "Bulldog" Walker, whose 8th Army was holding the perimeter at Daegu, coalition forces shoved the Korean People's Army (KPA) of the North all the way back, past the 38th Parallel (commonly referred to today as the DMZ) and nearly to the Chinese border at the Yalu River. This triggered a response from Chinese dictator Mao Tse Tung, who mobilized the Chinese People's Volunteers—a.k.a. Chinese Communist Forces (CCF)—which poured across the Yalu like a tsunami, catching the U.S. off guard. The result was a hasty allied retreat.

The combined forces of the KPA and CCF—the Chicoms—beat coalition forces back across the 38th parallel, and once again captured the South's capital of Seoul.

General Douglas MacArthur, who had conquered the Pacific less than a decade prior and had masterminded the amphibious assault of Incheon, was incensed at China's involvement. He began calling for operations to attack salient positions within China's borders. These operations included the use of nuclear weapons. President Truman made it clear that neither would happen. Fearing that an attack of China, within Chinese borders, would spark a world-ending nuclear holocaust, Truman ordered MacArthur, who had regained the initiative and was once again beating Chicom forces back, to halt at the previously established, albeit arbitrarily selected, 38th parallel, the border between the north and the south.

MacArthur, now harboring ire towards the President of the United States, ignored the order and continued to push north of the 38th, all the while publicly criticizing Truman whose next action was decisive. He relieved MacArthur of his command and replaced him with General Matthew Ridgeway. At this point, the war ground to a stalemate. Both sides agreed to seek peace but also "agreed that hostilities should continue even while negotiations were in progress."[1]   As a result, both sides began launching limited offensive operations to secure salient points across the 38th parallel, and the United States led coalition forces seeking to "strengthen the UN position at the negotiating table."[2] This dynamic would later become known as "limited war."

The Korean War is an outlier within the context of American wars. It is often referred to as the "forgotten war" by those who did not participate in it. Many today, especially among the millennial and post-millennial generations, are not even aware that it is known as "forgotten." Many of the aforementioned generations struggle to provide an answer when asked when or why the Korean War was fought. They can sometimes provide a rough time frame, generally answering in the form of a question, such as "in the 50's?" One in particular, when asked, stated: "I think it was in the 1800's." As far as why it was fought, the common answer is, "I have no idea."

Those born earlier, Generation X and Baby Boomers, typically provide more accurate responses, such as "between 1950 and 1954..." and "to prevent the spread of com-

---

1. T.R. Fehrenbach, *This Kind of War: The Classic Military History of the Korean War* (Open Road Integrated Media: New York, NY) 2014, p. 505.

2. John Toland, *In Mortal Combat: Korea 1950 - 1953,"* (Open Road Integrated Media: New York, NY) 2016, p. 450.

munism," respectively. Notwithstanding, this text is not meant to shed light upon this dilemma or cast dispersion at particular generations, although younger Americans having little to no knowledge about the Korean War is certainly concerning. It is not necessarily a result of their own failures, but of society at large. Although many texts have been written explaining the events that transpired, both prior and during the war, the war itself seemingly takes a back seat to learning about the World Wars and those wars that followed Korea. Of those texts written, few reference snipers. Those that do, more often than not, only do so in order to identify individuals who happened to be carrying a scoped rifle or happened to fire a shot at longer than average ranges. Fewer still truly outline the contributions of American snipers in the Korean War.

Unfortunately, in addition to overtly earning its moniker, many of the subtleties generated by the war, such as the recollections of individual American snipers, have also been forgotten. Whether that was by accident or by design is most certainly predicated on the disposition of those who lived it. Some may have wanted to forget while others may have struggled to find someone willing to listen. Nonetheless, anecdotal evidence of the American sniper's experience in the Korean War, much like World War II, is extremely rare and difficult to find. It is a certainty that sniper rifles that were designed by U.S. ordnance, less than a decade prior, existed and it is a certainty that these rifles saw use; however, most references supplied by primary sources almost always refers to the use of snipers by the Korean People's Army (North Korean) and the CCF, the two armies collectively known as "Chicoms." There are a small number of accounts of U.S. and allied troops utilizing scoped precision weapon systems early in the conflict, but the term "sniper" only loosely fits. If it was applied, it was in reference to any U.S. or allied military personnel who carried a scoped weapon.

That is not meant to diminish the contribution of those who did. Often enough, it was not they—but their commanders—who made the decision. It is also not mean to imply that those who trained themselves are less deserving of the title. Truly it is impressive that they were able to train themselves to utilize the weapon systems. Learning how to make "on-the-fly" calculations of elevation and windage adjustments, especially without formal training, is no small accomplishment and certainly merits the prestige of the now coveted label. It simply means that not all who carry scoped rifles are snipers, and not all snipers carry scoped rifles. Unfortunately, not all historians and/or war correspondents understand the subtle differences, nor should they be required to.

On the other side of the battle lines, the Chicoms were making excellent use of Soviet supplied Mosin-Nagant rifles equipped with PU and PE scopes, the same kind used by Zaitsev and Pavlychenko at Stalingrad during World War II. What made things worse is that this enemy's sniper threat was going largely unanswered as the U.S. and coalition forces in Korea had no way to reciprocate. Thus, as the Korean War was being fought, the U.S. once again found itself scrambling to identify and fill an empty void.

Subsequent to World War II, few snipers remained in service within the U.S. military. Of those that did, many preferred the same Springfield M1903 rifles that had been utilized in World War I. In addition to this, leadership had also begun mass producing new sniper variant Garands. These were assigned the designation M1D. They were "cheaper and faster to mass-produce than the C, and it utilized mostly standard parts but with the addition of a special heavyweight barrel."[3]

Still, the leaders of the U.S. military are nothing if not pragmatists. In the summer of 1952, after the initial back and forth of the war, when the commanding officer of the 3rd Battalion, 1st Marines, Lieutenant Colonel Thomas L. Ridge, had binoculars shot from his hands while gazing through them, he immediately called for U.S. snipers to find and eliminate enemy snipers. He was then informed, much to his chagrin, that there were no U.S. snipers to speak of. His response that followed was simple, "It's a helluva situation when the CO can't even take a look at the terrain he's defending without getting shot at! Something has got to be done about those goddamned snipers." He instructed his gunnery sergeants to "select and train a sniping squad as soon as it could possibly be organized."[4] Ridge would later receive a Silver Star for his actions in Korea, the citation reading, "Observing that hostile forces were stubbornly resisting the forward movement of his battalion, Lieutenant Colonel Ridge fearlessly moved elements of his command post to the immediate vicinity of the front lines in order to keep abreast of the situation and, repeatedly braving heavy hostile sniper and machine gun fire, skillfully directed his battalion's operations..."[5]

Other units throughout the Marines and the Army followed suit; however, much like World War II, no formalized or accredited program was developed, much less implemented. It is a certainty that during the Korean War, a number of mid-grade and

---

3. Pegler, *Out of Nowhere,* 273.

4. Ibid., 274.

5. "Thomas Ridge - Recipient," The Hall of Valor Project, accessed August 4, 2022.

higher-ranking officers believed the skillset to be necessary, one of which was Captain William S. Brophy who "made it his mission in Korea to convince the Army, at every level, that they should be teaching, training, and supplying their own snipers with the best instructions and weapons available."[6] Another, Colonel Henry E. Kelly, later authored an entire report titled, "Snipers, We Need Them Again"; however, his report has been entirely overlooked and now rests on a shelf in the archives at the U.S. Army War College Library at Carlisle Barracks, Pennsylvania.[7] That is not to say that developments within the sniper skillset were not made during this time frame. Although it is ultimately generals and politicians that establish formalized training programs within the military, formalized training programs are not solely responsible for the dissemination of information. That was as true then as it is now, and as the colloquialism states, "Necessity is the mother of invention."

A tried-and-true tactic that is utilized by modern snipers today is known as a "simo-shot." A simo-shot is a dynamic whereby two snipers will fire at a single pre-selected target at the same time. This accomplishes two goals. First, it increases the likelihood of a successful engagement. Second, it prevents the enemy from zeroing in on a single muzzle blast, thereby generating confusion and chaos within the enemy rank and file. This tactic is referenced in an X Corps "Combat Note" that outlined the techniques being successfully utilized by the 1st Marine Division and was being circulated through United Nation forces during the Korean War. In this way, despite the lack of large-scale training programs, successful snipers were able to relay valuable tactics to other snipers throughout allied lines.[8]

In an article titled, "They Call Their Shots," in the April 1953 edition of the *Marine Corps Gazette*, Lieutenant Colonel Glen E. Martin briefly outlined this trial-by-fire method of developing new skills when he wrote, "In surveying our sniper platoon program we found that some of our theories worked, others were workable with minor modifications, and some just wouldn't hold water. As time went on some of the snipers

6. Pegler, *Sniper: A History of the US Marksman.*, 147.

7. Michael Lee. Lanning, *Inside the Crosshairs: Snipers in Vietnam* (New York: Ballentine, 1998), 82.

8. Peter R. Senich, *U.S. Marine Corps Scout-Sniper: World War II and Korea* (Boulder, co: Paladin Press, 1993), 208.

came up with innovations of their own which were better than some of our carefully planned theories."[9]

Nevertheless, apart from a small number of commanders who acknowledged the combat multiplier benefit, and despite the recognition that a sniper weapon platform had been needed and created, this concept was lost on the United States military at large. It seems strange, almost derelict, that the military machine did not establish a permanent sniper program after lessons learned in World Wars I and II, especially when considering the sniper's overall contributions.

Perhaps it has been most eloquently worded by authors Charles Sasser, a former Army Green Beret and Craig Roberts, a former Marine Corps sniper, in their book *One Shot – One Kill*, when they wrote:

> "One hard-shooting sniper is worth more than a pair of light machine gun crews. His killing power is extensive and ubiquitous, plus the psychological factor of the single accurate shot cracking from nowhere lowers the enemy's morale and makes him afraid to take chances... nothing the enemy does goes unnoticed when four good snipers waiting for a target lie on a battalion's front observing the enemy through scopes."[10]

This sentiment is especially true considering that the terrain of the Korean peninsula is so well suited to the sniper's craft. The topography consists of mountainous terrain, which forced all transport to use narrow passes in order to move mechanized war implements and supplies. For the United States, these narrow passes were also utilized to facilitate troop movement en masse. That was not the case for the Chicoms, at least not on the front lines. The Chicoms, and more specifically, the CCF, would flood over the terrain on foot, marching over mountain as opposed to through the valley. They would occupy the high ground, in many cases, behind coalition lines; at which point, they would open up with hellish fire from mountain tops as allied troops retreated, forcing them into a gauntlet of death.

Once the war settled into what would become a two-year grind, the only reason for attacking a nearby hill was to benefit the respective side at the negotiating table. In this, a

---

9. Glen E. Martin, "They Call Their Shots," *Marine Corps Gazette*, April 1953, pp. 25-28.

10. Sasser and Roberts, *One Shot – One Kill*, 82.

paradigm shift occurred. It would be the beginnings of "limited-war" doctrine, wherein a sniper could have been especially effective. Indeed, what was taking place was not lost on the troops. They knew that they were not taking hills to conquer a foe, but rather to expedite a treaty. One can only imagine the subsequent deficit in morale knowing that a total victory, such as that which was accomplished only a few years earlier at the end of World War II, was a forgone conclusion. It must have been painful to come to terms with the idea that sacrifices—to include that of life—were being made upon the alter of compromise at best. Neither the Army, nor the Marines, were prosecuting a war of maneuver, but of occupation, more similar to a widespread medieval siege than a modern conflict. What is perhaps more detrimental is that the U.S. troops knew one thing for certain, they were not allowed to win.

Throughout the conflict, the Ordnance Department had been working on a few "wonder weapons." Namely, napalm, which saw its first use at the end of World War II. Furthermore, napalm bombs, in effect, proved second in destruction only to nuclear weapons. Though civilian populations may have feared annihilation via nuclear holocaust, it is arguable that the dread produced in the enemy by the specter of burning to death due to napalm was unsurpassed. In any event, napalm worked so well in Korea that it saw unprecedented use in the next major war in which the U.S. would participate. It is in this revelation that a familiar question arises: Why was mass destruction wrought from the air tacitly accepted while the precise application of a single bullet was explicitly scorned? After all, both generated immense anxiety within the enemy. Both originated from afar, but only one killed via melting flesh from the bone.

Nevertheless, the concept of a permanent sniper program was not lost on the entire U.S. military. The U.S. Army Marksmanship Unit (AMU) attempted an effort at establishing a formalized sniper training program at Camp Perry, Ohio shortly after the Korean War came to an armistice. Unfortunately, the program was short lived and only lasted one year, from 1955 to 1956. In retrospect, it seems counterintuitive, with battle sight names like "Sniper Ridge," that the U.S. did not try harder to establish a permanent formalized program after the Korean War, especially considering the Korean peninsula was ideally suited for sniper craft. There are those who would argue that the sniper role was still not entirely understood; however, that seems like a measure to excuse a flagrant oversight. Notwithstanding, much like the wars preceding, when the armistice that formally ended, the fighting between North and South Korea was signed, those who had earned the title "sniper," via trial by fire no less, stowed their rifles and returned home ready to forget.

NOTES:

1. T.R. Fehrenbach, *This Kind of War: The Classic Military History of the Korean War* (Open Road Integrated Media: New York, NY) 2014, p. 505.

2. John Toland, *In Mortal Combat: Korea 1950 - 1953,"* (Open Road Integrated Media: New York, NY) 2016, p. 450.

3. Pegler, *Out of Nowhere,* 273.

4. Pegler, *Out of Nowhere,* 274.

5. "Thomas Ridge - Recipient," The Hall of Valor Project, accessed August 4, 2022.

6. Pegler, *Sniper: A History of the US Marksman.*, 147.

7. Michael Lee. Lanning, *Inside the Crosshairs: Snipers in Vietnam* (New York: Ballentine, 1998), 82.

8. Peter R. Senich, *U.S. Marine Corps Scout-Sniper: World War II and Korea* (Boulder, co: Paladin Press, 1993), 208.

9. Glen E. Martin, "They Call Their Shots," *Marine Corps Gazette*, April 1953, pp. 25-28.

10. Sasser and Roberts, *One Shot – One Kill,* 82.

CHAPTER VI

THE VIETNAM WAR

Whereas World War II may be the most studied conflict in American history, Vietnam is easily the second. Much like Korea a little over a decade prior, most Americans could not have pointed to Vietnam on a map prior to the United States becoming involved there. Insofar as American interests are concerned, Vietnam had little to offer. Nevertheless, as President Eisenhower left office in 1961 to be succeeded by President John F. Kennedy, the spread of communism in the Far East was generating dread within the U.S. government.

The devastating impacts of World War II had widespread effects. No single nation was left untouched. In particular, it had sapped France's ability to govern its colonial interests to include that of French Indochina, what is today the countries of Vietnam, Cambodia, and Laos. This created a power vacuum in the area and, though it revitalized a sense of nationalism within the respective indigenous populations, it also created a path whereby communist influence could encroach. In 1949, four years after the Japanese surrendered to the United States officially ending the second World War, Chinese communist forces began supplying Ho Chi Minh, the spiritual, political, and military leader of the Viet Minh forces of northern Vietnam, with Soviet and Chinese weapons.

Ho had risen to power and taken control of the northern city of Hanoi subsequent to the 1945 treaty that saw the expulsion of the Japanese from Vietnam. In 1954, with Soviet and Chinese support, he was able to defeat the French at Dien Bien Phu, bringing

France's colonial rule of Indochina to a permanent end. The cessation of hostilities was negotiated and a treaty signed in Geneva, Switzerland later that year. Similar to what had taken place on the Korean peninsula, a boundary line was arbitrarily selected and drawn, splitting the country at the 17th parallel.

Ho became president of the communist-backed north, and with French support, Emperor Bao Dai ruled in the south. Emperor Bao's rule, however, would be short lived. A coup in the south in 1955 saw a fiercely anti-communist politician named Ngo Dinh Diem usurp power from Emperor Bao, naming himself President of what would become the Republic of Vietnam. Although the United States had little interest in securing the colonial interests of France, President Eisenhower saw an opportunity in President Diem to halt a proverbial communist domino from falling in Southeast Asia and pledged American support to Diem. Nevertheless, the Korean armistice had been signed only two years prior, and the people of the United States had zero appetite for another war. At the end of 1955, in favor of preventing the United States from entering into another war, Eisenhower instead deployed a Military Assistance Advisory Group, which consisted of U.S. Military personnel, in order to train the newly created Army of the Republic of Vietnam, or ARVN.

In 1962, following President Kennedy's 1961 election, the Military Assistance Command Vietnam (MAC-V) was created. U.S. Army Special Forces soldiers were then deployed to Vietnam to continue the training of ARVN troops. Unfortunately, President Diem's regime was plagued by corruption and on November 2, 1963, a military coup d'état had taken place within South Vietnam. It resulted in the summary execution of President Diem on a roadside in Saigon, present day Ho Chi Minh City. It is clear that President Kennedy's administration, having negotiated myriad obstacles generated by the Diem administration's corruption, tacitly approved of the coup. However, what is not clear is whether President Kennedy had sanctioned the assassination. Those close to the president indicated that he appeared genuinely distraught upon learning of Diem's murder, indicating that the U.S. President was not aware of the plan to kill Diem. While speculation exists for both sides of the argument, a definitive answer may never be known as President Kennedy was assassinated in Dallas, Texas just twenty days later. During Kennedy's administration, he had consistently increased the number of MAC-V personnel in Vietnam. On November 22, 1963, when President Lyndon B. Johnson was sworn in on board Air Force One, there were some 16,000 U.S. military advisors in South Vietnam.[1]

The persistent build-up of U.S. military personnel in South Vietnam, regardless of their officially sanctioned capacity, meant that it was only a matter of time before mounting tensions would erupt into combat between U.S. personnel and the North Vietnamese Army. To be clear, at this point Americans had engaged with the enemy, and Americans had been killed in Vietnam, but these occasions were somewhat few and far between. This changed; however, when two U.S. Navy destroyer class ships were attacked, between August 2 and 4, 1964, in the Gulf of Tonkin, off the east coast of North Vietnam. The two ships, the *USS Maddox* and the *USS Turner Joy*, reported that they had been attacked by North Vietnamese torpedo boats and returned fire. There is, presently, speculation as to whether the event actually took place. Whether it did or not, President Johnson and Secretary of Defense Robert McNamara did not let the news of the exchange, fabricated or otherwise, go to waste. After reports of the second attack on August 4, President Johnson went to the United States Congress and the Gulf of Tonkin Resolution was passed. Of the 535 members of Congress, 533 supported the resolution, the only two opposing votes having been cast within the Senate.[2] The resolution gave President Johnson, and later President Nixon, broad war making powers. The decision was made to become directly involved in the ongoing conflict in Vietnam, and Operation Rolling Thunder was launched, a bombing campaign of North Vietnam and neighboring Laos, which had by then been invaded and was under the control of North Vietnam,. What would follow was an eight-year, controversy generating, slug fest during which the terms "limited war," "seek and destroy," and "hearts and minds" would become common place. Nevertheless, the cacophony of war within the mountains and jungles of Vietnam would also be the birth cry of the modern-day professional sniper.

In March of 1965, when the first U.S. combat troops entered the country of South Vietnam, they found themselves involved in an unfamiliar warfare doctrine. Although the North Vietnamese Army was a uniformed combatant, North Vietnamese sympathizers in the south began launching guerrilla style attacks against U.S. forces and their counterparts, the ARVN. These sympathizers, derisively labeled Viet Cong or "VC" for short,

1. "Military Advisors in Vietnam: 1963," Military Advisors in Vietnam: 1963 | JFK Library, accessed August 19, 2022.

2. "U.S. Involvement in the Vietnam War: The Gulf of Tonkin and Escalation, 1964," U.S. Department of State (U.S. Department of State), accessed August 19, 2022.

were ununiformed and utilized hit-and-run-style tactics. In addition to this, the majority of the routes that supplied the VC and NVA troops in the south ran through Laos and Cambodia, countries that the U.S. was forbidden from occupying. This supply line, known as the Ho Chi Minh Trail, proved to be nearly impossible to disrupt. Although bombing of the trail within the Laos and Cambodian borders was permitted, without ground forces to take and hold the conquered territory, the trail was merely repaired or rerouted, sometimes within the space of a few hours. In order to stifle the guerrilla style attacks, General William Westmoreland implemented a "search and destroy" doctrine, whereby troops would seek out and eliminate pockets of NVA and/or VC throughout South Vietnam. The tactic proved wholly inadequate as the NVA and VC were adept at displacing. In some circumstances, as was the case at Hill 937 in the A Shau Valley, infamously remembered as "Hamburger Hill..." U.S. troops battled for days to take the hilltop only to abandon it almost immediately afterwards. Enemy troops that had retreated merely reoccupied it after U.S. forces left.

The U.S., in concert with the Republic of Vietnam government, focused instead on a "hearts and minds" campaign by introducing the Strategic Hamlet Program. The goal was to win the hearts and minds of the people of South Vietnam by providing supplies, jobs, and homes to the people occupying rural areas in South Vietnam. The governments reasoned that providing a system of welfare and infrastructure would, in turn, win support for democracy within the south and possibly convert those who were on the fence, so to speak, about communist ideology. What it succeeded in doing was creating checkpoints at which VC operating in the area could be resupplied by non-combatant sympathizers, or, in some cases, locations where VC could pillage resources that had been supplied to the indigenous population by the U.S. and South Vietnamese governments. Needless to say, it was a disaster. Those who lived in the hamlets lived in hell, constantly held in contempt by combatants on both sides, always suspected of aiding and abetting the enemy, whether they were or not. Although the strategic hamlet program appeared altruistic on paper, in practice it left much to be desired. In any event, the overall "why" and "how" the Vietnam war was fought is not the focus of this text. Whether the United States should have been involved there is for other historians to debate. What is certain is that no previous American war so clearly defined the role of the American military sniper.

Throughout the course of their training, modern American snipers cut their teeth on the stories of legends born in the jungles of Southeast Asia. The Vietnam War proved to be the paradigm shift insofar as understanding what a sniper was capable of accomplishing.

It would eventually lead to the establishment of permanent sniper programs within the U.S. military, beginning with the United States Marine Corps. But that would not come until later. Like every war preceding it, U.S. forces once again constructed ad hoc sniper schools in-theater to train troops in the midst of ongoing combat. Within the Marine Corps, this task was undertaken by a young officer named Captain E. J. Land and a young NCO named Staff Sergeant Carlos Hathcock II. Following orders from the 1st Marine Division Commanding General, the two began training snipers at Hill 55, named Camp Muir at that time, approximately nine miles southwest of Da Nang in South Vietnam.[3] Hathcock, who had won the 1,000-yard shooting contest known as the Wimbledon Cup, and his commander, Land, set to work training snipers as quickly as they could. Hathcock, one of the aforementioned legends, did not become so, however, by training snipers.

Indeed, no text detailing snipers would be complete without mention of Carlos Hathcock II. He was born in Little Rock, Arkansas in May of 1942 and was well acquainted with rifles by the time he was a teenager. He enlisted in the United States Marine Corps in 1959, when he was 17 years old. In 1966, Hathcock was deployed to Vietnam where he quickly found his way into a sniper's role. He is recognized as one of the most legendary snipers to have ever wielded a rifle in American military history and is also, arguably, the individual responsible for the shift, albeit slight, in soldier and civilian perception of snipers. That shift, however, would not take place until some two and a half decades after his exploits. He is credited with 93 confirmed kills, which is no small accomplishment, but his legend is not predicated on his kill count entirely as there were snipers in Vietnam who were credited with more confirmed kills.

Thus, the question must be asked why his name is instantly recognizable, at least by snipers, and the names of other snipers, with similar kill counts, are not instantly recognized? After all, Charles "Chuck" Mawhinney, who was also in the United States Marine Corps and served in the Vietnam War, is credited with 103 confirmed kills. Another, Adelbert "Bert" Waldron, a sniper in the U.S. Army's 9th Infantry Division operating in the Mekong Delta of South Vietnam is credited with between 109 and 113 confirmed kills, second only to Chris Kyle, a U.S. Navy Seal sniper credited with 160 kills during the twenty-first century wars in Iraq and Afghanistan.

---

3. Charles Henderson, *Marine Sniper: 93 Confirmed Kills* (New York: Berkley Caliber Books, 2005), 101.

In addition to that, Bert engaged and eliminated—from a moving boat—an enemy sniper who was hidden in a tree some 900 yards away. The answer to this question will no doubt ruffle feathers, especially among Marine Corps Scout Snipers, both past and present. The legend of Carlos "White Feather" Hathcock was born of and largely predicated on his own recollection. Though that seems somewhat provocative, it is in no way meant to diminish his accomplishments or contributions. Many of Carlos' deeds were witnessed, removing all doubt as to whether or not they occurred. Nevertheless, there exists no empirical evidence of some of his deeds, which in some cases are so fantastical that a level of faith typically reserved for religious belief must be utilized in order to remove all doubt of their validity.

One event recounted by Carlos was his counter-sniper operation against a North Vietnamese sniper who had been sent specifically to kill him. During Carlos' stalk, the enemy sniper fired at him, but by chance, Carlos had tripped over a fallen tree and the round missed, striking Carlos' spotter's canteen. The enemy sniper, whom is identified only with the sobriquet of "the Cobra," displaced to find another firing position. Carlos stated that the enemy sniper relocated to a sub-nominal position, facing into the sun. He claimed that he saw the sun glint on the enemy sniper's objective lens, turned, aimed, and fired at where he saw the glint of light and sent the bullet through the enemy sniper's scope and through his eye socket. That said, it is a fact that Carlos did go on missions during which he was stalking enemy snipers, and it is undisputed that Carlos found and eliminated enemy snipers. Whether or not Carlos killed the Cobra via shooting straight down the center of his scope; however, is a matter that may never be proven beyond a doubt. Considering the variables involved, to include the components utilized for the 30-06 ammunition utilized by Hathcock in his Winchester Model 70 rifle, and the design of the Soviet-made PU and PE scopes utilized by the NVA during that time frame, as well as a number of other factors, to include the distance the projectile traveled—which would dictate the parabolic arc, or trajectory, that the bullet traveled—and the velocity at which it reached its intended target, it is highly unlikely that the event occurred just as Carlos recounted it. On the other hand, it may very well have. The intent of this text is not to prove that one individual sniper did or did not accomplish such a feat, but rather to identify the impacts that individual snipers had, positive or negative, on the discipline throughout the course of American history. At risk of a Machiavellian claim that "the ends justify the means," the argument could be made that it does not matter whether the event occurred just as Carlos *claimed* it did because the overall impact the claim had on the science of the long gun could be characterized as positive.

Nevertheless, this particular deed, in addition to his others, would not become well-known feats until 1993, when he did his one and only interview with Major John L. Plaster, himself considered a prolific sniper. Plaster published the book *The Ultimate Sniper: An Advanced Training Manual for Military and Police Snipers,* and subsequently published a video to accompany the text. It proved to be the beginning of an era during which the sniper became the hero. This particular display of sniper prowess, whether fabricated or not, created a pop-culture phenomenon and has been replicated in nearly every sniper novel, movie, and video game since.

The feature film *Saving Private Ryan* features a sniper character played by Barry Pepper who engages and eliminates a Nazi sniper via shooting him through the scope and through the eye. The films *Sniper* starring Tom Berenger, *Shooter* starring Mark Wahlberg, and even *American Sniper* starring Bradley Cooper all feature the same feat. Indeed, any story or film featuring a sniper would not be complete without it. In any event, the reason Carlos Hathcock became so recognized is because he was one of the first, if not—the first—to discuss his role as a sniper in warfare in a manner by which he demonstrated pride in his service to his country and pride in the discipline to which he had dedicated himself. Nevertheless, as previously mentioned, these changes would not come about until long after combat soldiers had left Vietnam. At that time, in the northern portion of South Vietnam, near the 17th parallel and the A Shau Valley, despite their harrowing missions and tremendous contributions, unless they were saving the day, snipers were still held in contempt.

Meanwhile, Army line commanders in the south near Saigon also recognized the need for snipers and set to the task as well. Once again, the Army Marksmanship Unit was recruited to train U.S. Army snipers, and a school was constructed near the Mekong Delta in South Vietnam.[4] The Army school focused on training 9th Infantry Division soldiers to become snipers. During this period of time, a small change was made within the 9th Infantry Division in the U.S. Army that would have a resounding impact on the sniper programs of the U.S. military.

Prior to this, as was the case in both World Wars and to some extent, the Korean War, snipers, much like any other line infantryman, reported directly to their respective

---

4. Pegler, *Sniper: A History of the US Marksman,* 157.

company commanders. The problem was that few company commanders understood how to facilitate the use of a sniper. In many instances, snipers were utilized in the same manner as any other infantry soldier, being put on guard duty or other day-to-day infantry tasks. Within the 9th Infantry Division in 1968, however, command of snipers was pulled from company commanders and assigned to battalion commanders.[5] This prevented snipers from being tasked to non-role specific duties, allowing them to focus on producing results. Instead of large infantry platoons patrolling the jungle with a sniper within one of the fire-teams, small teams of snipers, with three or four infantrymen for security, could move stealthily through the jungle and provide support, eliminate targets, or conduct counter sniper operations.

Despite the new energy being devoted to training and fielding snipers and the investment into designated sniper weapon systems, old and familiar ways of thinking still plagued the profession. As brilliantly stated by Colonel Michael Lee Lanning, a commander during the Vietnam War, "American bombardiers released tons of bombs on enemy soldiers and innocent civilians alike from 20,000 feet. Artillerymen fired large-caliber guns with no warning at targets miles away. Anonymous killing at a distance was acceptable—unless it was by an individual marksman sighting his individual enemy through a telescopic sight and squeezing the trigger to fire a single deadly bullet."[6] Much like all the wars prior, snipers were considered a "necessary evil." As a result, derisive titles such as "Murder Inc." were developed to characterize snipers.

One Marine sniper and veteran of the Vietnam War, John Culbertson, titled his memoir with one of the pejorative epithets: *13 Cent Killers: The 5th Marine Snipers in Vietnam.*[7] It is arguable that pulling snipers away from line commander's authority and having them report, instead, to battalion commanders did not help the social dilemma. As a result of the disdain, snipers often kept to themselves anyway. When they did, on occasion, open up, it was typically to other snipers. Whether they wanted to sequester themselves or not, they did. Largely they did so for survival. Not survival in a literal sense, but figuratively—mentally. To be hated by your enemy is expected. To be hated by your peers is not. As is the nature of humanity, the less something is understood, the easier

---

5. Ira Augustus Hunt, *The 9th Infantry Division in Vietnam: Unparalleled and Unequaled* (Lexington, KY: University Press of Kentucky, 2010), 100.

6. Lanning, *Inside the Crosshairs*, 78.

7. John J. Culbertson, *13 Cent Killers: the 5th Marine Snipers in Vietnam* (New York: Ballantine Books, 2003).

it is to hate. Snipers were caught in an idiomatic Catch-22. In Vietnam they ostracized themselves to escape the hatred of their peers who hated them even more for ostracizing themselves.

Despite the good work, lives saved, and other accomplishments by the various snipers in Vietnam, at the end of the war, sniper programs were once again abandoned. This time, however, leadership within the programs would not go down quietly. Specifically, Captain Land, who had since been promoted to Major, and Staff Sergeant Hathcock, who had since been promoted to Gunnery Sergeant, began petitioning the Pentagon.

Land and Hathcock succeeded in proving their case, and a permanent sniper school, the Marine Scout/Sniper School, was established in Quantico, Virginia. Since then, the Marine Schools of Infantry East at Camp Geiger, North Carolina and West, at Camp Pendleton, California, have both established Scout/Sniper training programs. The U.S. Army, unfortunately, did abandon their program until 1987 when a permanent Sniper School was established at the School of Infantry in Fort Benning, Georgia.

NOTES:

1. "Military Advisors in Vietnam: 1963," Military Advisors in Vietnam: 1963 | JFK Library, accessed August 19, 2022.

2. "U.S. Involvement in the Vietnam War: The Gulf of Tonkin and Escalation, 1964," U.S. Department of State (U.S. Department of State), accessed August 19, 2022.

3. Charles Henderson, *Marine Sniper: 93 Confirmed Kills* (New York: Berkley Caliber Books, 2005), 101.

4. Pegler, *Sniper: A History of the US Marksman,* 157.

5. Ira Augustus Hunt, *The 9th Infantry Division in Vietnam: Unparalleled and Unequaled* (Lexington, KY: University Press of Kentucky, 2010), 100.

6. Lanning, *Inside the Crosshairs*, 78.

7. John J. Culbertson, *13 Cent Killers: the 5th Marine Snipers in Vietnam* (New York: Ballantine Books, 2003).

PART IV

Modern Conflict

"We killed the bad guys and brought the leaders to the peace table. That is how the world works."

-Chris Kyle

CHAPTER VII

THE PERSIAN GULF AND MOGADISHU, SOMALIA

By the end of 1989, the Cold War had come to a close. The Peaceful Revolution, as it became known, saw the destruction of the Berlin Wall, separating East and West Germany. By 1991 the Soviet Union had dissolved, and it seemed as though the threat of communism, or its proliferation at least, would pass into the history books. The world breathed a collective sigh of relief, no longer having to remain hyper-focused on a nuclear holocaust brought on by two superpowers at war.

Unfortunately, a new threat was developing. There is a tendency to attribute the eruption of war to a single isolated incident. Typically, those incidents make headlines on the world stage. The wake of destruction that follows overshadows the multitude of variables included in the chain of events that escalated into the conflict. For example, many attribute the eruption of World War I entirely to the assassination of Archduke Ferdinand. After all, it seems a plausible, and acceptable, reason for war. Nevertheless, depending upon one's approach to the subject, it is arguable that World War I began its incubation when Kaiser Wilhelm of Germany sought to isolate France on the geopolitical stage, some four decades earlier.

The same could be said of every conflict the United States has ever been involved in and including the first Persian Gulf War. In the early 1990s, when the United States became involved in Operations Desert Shield and Desert Storm, also known as the Gulf War, the single headline making justification was Iraq's invasion of Kuwait and Saudi

Arabia. That is indeed the cited reason for the 35-nation coalition led by the United States' involvement. However, what must be recognized is that this particular part of the world is responsible for the vast majority of the world's oil production and distribution and that Kuwait, despite its small size, was, and remains, an oil rich nation. Destabilization in this region would have had a devastating impact, economically, on the world stage. On the other hand, control of this region would guarantee wealth and power globally. When the President of Iraq, Saddam Hussein, ordered the invasion of Kuwait in August of 1990, this was, perhaps, his ultimate goal.

Saddam reasoned that by conquering Kuwait he would not only eliminate the large amount of debt that Iraq owed to the tiny nation, but that the control of its oil fields would enrich Iraq and increase its influence on the region. After all, Saddam needed both after the protracted Iraq Iran war fought from 1980 to 1988. Kuwait fell to Iraqi forces quickly; at that point, Saddam turned his sights towards Saudi Arabia. King Fahd of Saudi Arabia, concerned with the Iraqi troops amassing on his border with Kuwait, then invited western forces into Saudi Arabia to assist in the defense of the kingdom. This invitation to westerners to help in the defense of Saudi Arabia, which happens to be home to two of the three holiest sites in the Islamic faith, the Masjid al-Haram (Grand Mosque) in Mecca and the Masjid an-Nabawi (Prophet's Mosque) in Medina, enraged some within the Islamic world. A militant group in the region that was headquartered in Afghanistan and known simply as The Base, or *al-Qaeda*, offered its services to King Fahd, who rejected the offer.

The founder of the group, a man who would later be recorded infamously in history, was Osama Bin Laden. Bin Laden was incensed that King Fahd would choose *infidels*, or those who were not of Islamic faith, to protect Saudi Arabia over his own Muslim fighters, especially after those same fighters had proven themselves capable during Russia's occupancy of Afghanistan. He turned his hateful gaze towards the United States for the perceived slight and began plotting revenge which would later manifest as attacks on the World Trade Center in New York City in 1993 and again in 2001. Meanwhile, as coalition forces arrived in the Kingdom of Saud, they were provided with a clear objective; stop the Iraqi invasion of Saudi Arabia and push Iraqi forces out of Kuwait.

By this time, the sniper schools that existed within the U.S. military had been producing some of the most elite shooters the world had ever seen. Furthermore, investments had been made in the tools of the trade. During and after the war in Vietnam, dramatic shifts had taken place insofar as weapon platforms and calibers. Across the branches, the venerated M14 had been replaced by Eugene Stoner's AR-15 platform rifle and was designated the M16. Carlos Hathcock's Winchester Model 70 chambered in 30/06 had been replaced by Remington's offering, the Model 700, chambered in 7.62 x 51 NATO.

In 1988, within the U.S. Army, the M-24 Sniper Weapon System was issued to snipers. It featured an H-S Precision synthetic molded stock with an adjustable length of pull; the Leupold Ultra Mark 3, 10 x 42mm fixed power scope with a mil-dot reticle; and a Remington Model 700 Long-Action. Although the rifle was initially chambered in .308 (7.62 x 51mm NATO.) Within the United States Marine Corps, the Scout Snipers had been utilizing the Remington 700 since 1966. Nevertheless, the original M40 featured a wooden stock. In 1977, the M40A1 began production and would remain the primary U.S.M.C. sniper rifle until 1996. The A1 variant, unlike its Army M24 counterpart, featured a Short-Action, a fiberglass McMillan stock, and a 10-power fixed Unertl scope with a mil-dot reticle developed by the Unertl company specifically for the Marine Corps. In the years since, these rifles and their optics have been upgraded in favor of technologically superior offerings, but during the Persian Gulf War, these were the tools that Army and Marine snipers carried to war.

The majority of the Persian Gulf War is remembered for its aerial and mechanized elements. General Norman Schwarzkopf, commander of the coalition forces in the middle east, devised a Jominian "war-of-maneuver" style campaign whereby coalition forces would confront enemy forces head on in Kuwait while a second element would flank from the left. The operation, known as the "Left Hook" would push deep into Iraq and cut off the enemy forces within Kuwait. Despite this mechanized warfare plan, snipers within the U.S. Army and Marine Corps were present and applying their trade.

An article in the *New York Times* dated February 3, 1991, provides an account from a unit of Marine Scout Snipers who were to deploy to the Kuwaiti border with Saudi Arabia the next morning. Although the article, written by Chris Hedges, was brief and only provides a glimpse into a day in the life of a sniper, it does a splendid job, perhaps unknowingly, of illustrating a number of dynamics to include the mindset of a sniper, a sniper's view on taking a life, typical sniper culture, the contempt with which snipers were held by fellow troops, and the reciprocation of that contempt for those who avoided military service.

In one paragraph, Hedges writes "The bonding of the unit, which will act as the eyes and ears of the battalion from forward positions, is tight. Platoon members are disdainful even of fellow marines, who in turn chafe at the snipers' arrogance."[1] Hedges incorrectly assumes that the antipathy displayed by the "other marines" came as a result of the snipers' arrogance. It is not unexpected as Hedges more than likely was unaware of the long history of derision between snipers and their non-sniper counterparts. He could not have known that what he termed "arrogance" was merely psychological walls erected in order to further defend sacred honor. Truly, it was defense mechanisms at work.

Perhaps the most well-known account of snipers in the first Persian Gulf war is that written by Anthony Swofford and published in 2003. His memoir *Jarhead* would go on to become a feature film of the same name starring Jake Gyllenhaal and Jamie Foxx. Swofford's account details his training and deployment to the Persian Gulf. Swafford's novel seemingly indicates that in an age of modern warfare, dominated by mechanization and airpower, snipers are somewhat superfluous.

His point is somewhat justified by his own experience. Throughout his deployment, he never fired his rifle despite having several opportunities to do so. One of those opportunities was during a mission to take Ahmed Al Jaber Airfield, some thirty miles southwest of Kuwait City. After arriving in position and setting up their hide, he and his spotter witnessed two Iraqi Republican Guard officers, one of them obviously high ranking, arguing with one another in the air traffic control tower of the airfield. He recalled that the two eventually became involved in a scuffle and were broken up by subordinates. He requested permission to engage the two HVTs, writing:

> The men in the tower are perfect targets. The windows are blown out
> of the tower, and the men are standing, and I know that I can make a
> headshot. Johnny has already called the dope for the shot. He thinks I can
> take two people out in succession, the commander who wants to fight and
> one of his lieutenants. He thinks that the remaining men in the tower will
> surrender plus however many men are under that command, perhaps the
> entire defensive posture at the airfield.[2]

---

1. Chris Hedges, "War in the Gulf: The Marines; War Is Vivid in the Gun Sights of a Sniper," *The New York Times*, February 3, 1991, p. 1.

2. Anthony Swofford, *Jarhead: A Marine's Chronicle of the Gulf War and Other Battles* (New York City, NY: Scribner, 2003), 240.

Swofford was denied permission to engage. The commander provided the reasoning that engaging and eliminating the Iraqi officers would induce a fight and eliminate the possibility of a surrender. The misinformed commander then utilized infantry to breach the compound and start a fight. Swofford was obviously disenchanted by this, surmising that, "The captains want some war, and they must know that the possibilities are dwindling..."[3] Today, the sniper's ability to eliminate an HVT and possibly force a mass surrender is better understood . . . though only slightly. Swofford's account is somewhat typical of what most operational snipers encounter on a day-to-day basis.

The amount of training that accompanies the skillset is nearly overwhelming, and the manner in which a sniper is trained has a permanent psychological impact. Thus, his disappointment in not being able to apply his skillset is understandable. Ruminating on the "what could have been" is mentally taxing, especially considering that he subsequently witnessed a battle that he knew he possibly could have prevented. However, the speculation that snipers are no longer necessary due to technological advances in warfare could not be further from the truth. Deadly accurate precision rifle fire aside, a sniper's primary function is to provide real-time intelligence. Despite never having squeezed the trigger on a target, Swofford and his spotter excelled at this aspect. Insofar as sniper's are concerned, there are relatively few published accounts of operational snipers outside of those already mentioned. Again, the sniper profession is clandestine in nature and was not featured within popular culture as often as it would later be.

The Persian Gulf War was swift and decisive. When it came to an end, the Iraqi army had been nearly wiped out. The final coalition attack occurred on the evening of February 26 and extended into the morning of February 27, 1991. It would later become known as the "Highway of Death."

Coalition aircraft from multiple nations to include the United States, Canada, Britain, and France effectively boxed in the retreating Iraqi military on Highway 80, leading from Kuwait City to Basra, Iraq. The resulting carnage became a matter of controversy, some claiming that the force utilized by coalition forces was excessive considering the Iraqi army was retreating. Regardless of one's opinion on the matter, it ensured that the Iraqi army would never again be able to initiate an unprovoked invasion of another sovereign nation. The Persian Gulf War came to an end the following day, on February 28, 1991.

---

3. Swofford, *Jarhead*, 241

Near the end of 1993, nearly three years after the end of the Persian Gulf War, the U.S. got its first taste of unconventional, non-linear, insurgent style warfare since Vietnam. The events unfolded in Somalia, a war-torn country that was ruled by warring factions of despotic tyrants. The constant chaos created an environment in which the innocent fell victim to various warlords. As a result, starvation was rampant. The warlords fought for, among other things, control of the food supply which in-turn guaranteed their power.

The United States Army arrived in the capital city of Mogadishu in 1992 on a humanitarian mission to distribute food; however, subsequent to their arrival "U.S. troops were slowly drawn into interclan power struggles and ill-defined 'nation-building' missions."[4] Within the sprawling city, one of these warlords, an individual named General Muhammad Farah Aideed, had begun targeting U.N. peacekeepers and on August 8, 1993, ordered the use of a command detonated mine that resulted in the deaths of four U.S. Military Police.[5]

From this point, the attacks against U.S. personnel only increased. As a result, the U.S. defense posture also increased to include direct actions to locate and apprehend those responsible for the attacks. One of these missions that took place on October 3 and 4 of 1993, known as Operation Gothic Serpent, was led by a United States Army unit designated Task Force Ranger. The mission's goal was to apprehend Aideed; however, the situation quickly deteriorated and turned, instead, into a rescue mission. The operation would be forever remembered as the "Blackhawk Down Incident."

Most are aware of the events that occurred during the course of those two days, especially since the release of the feature film *Black Hawk Down* featuring Josh Hartnett and Eric Bana. Some have even read the book, written by Mark Bowden, on which the film was based. During those fateful hours two U.S. Army Delta Force snipers, Sergeant First Class Randy Shughart and Master Sergeant Gary Gordon, proved that snipers are, and always will be, necessary.

Some incorrectly assume that snipers have a predilection towards killing. That is not the case. Some are fascinated by the science of the long gun. Others are fascinated by the stalk. Some just happen to have grown up shooting and display exceptional talent or skill. It is safe to say that all snipers hope that they can utilize their skillset to save lives. It is

4. General John S. Brown "Introduction," United States Forces, Somalia after Action Report and Historical Overview: The United States Army in Somalia, 1992-1994 § (2003).

5. "United States Forces, Somalia after Action Report and Historical Overview: The United States Army in Somalia, 1992-1994 § (2003).

also safe to assume that most—almost all—do not wish to take a life but will if necessary. As with anything else, exceptions obviously exist. Indeed, that is the nature of the human condition.

Nevertheless, it is the desire to save lives that drives most snipers. If taking one life saves ten, they will do it every time without a second thought. That is the empathy that sets a soldier apart, their willingness to sacrifice a part of themselves, or their self entirely, on the behalf of others. For snipers, this sacrifice can be far more intimate. When holding the crosshairs on a human being and squeezing the trigger, it is possible that the sniper may witness the last grimace on the face of the target as they take their last breath, or perhaps notice a wedding ring on their hand. It is no longer a distant silhouette without human features. In some cases, the sacrifice made may be far more than psychological. That was the case in the situation involving SFC Shughart and MSG Gordon in Mogadishu, Somalia on October 3, 1993.

At approximately 13:50 hours, the codeword "Irene" was called alerting U.S. forces to begin Operation Gothic Serpent, which would later be known as the Battle of Mogadishu. Approximately forty minutes after the operation began, the first Blackhawk Helicopter, call sign Super 61 (Super Six One,) was hit by an RPG-7 warhead. After Super 61 went down, the focus shifted to rescuing survivors of the crash. An MH-6 Little Bird helicopter quickly landed and collected survivors from Super 61, but only those who could move on their own. Within eight minutes, Blackhawk Super 68 (Super Six Eight) proceeded to the location of the downed helicopter and began offloading a Combat Search and Rescue (CSAR) team.

During the offload, Super 68, too, was struck by an RPG, but survived the impact. Super 68 finished the CSAR offload and subsequently returned to base. The CSAR team, after having roped to the ground, discovered that both the pilot, CW3 Cliff "Elvis" Wolcott, and co-pilot CW3 Donovan "Bull" Briley, had both been killed in the crash. While negotiating heavy enemy fire, the CSAR team extricated Briley and Wolcott as well as, the left door gunner, Sergeant Ray Dowdy, who was still alive.[6]

At approximately 16:40 hours, while providing air coverage of the wreckage of Super 61, Blackhawk Super 64 (Super Six Four) piloted by CW3 Michael Durant became the third helicopter to get hit by an RPG-7 warhead that day. The RPG damaged the tail rotor of the helicopter which held together for a short period of time before coming apart.

---

6. Mark Bowden, *Blackhawk Down: An American War Story* (Philadelphia, PA: Philadelphia Inquirer, 1997), 18.

As a result, not long after Super 61 went down, Super 64 became the second Blackhawk helicopter to crash. Durant sustained severe injuries to his spine during the crash. His three-man crew would all succumb to mortal injuries sustained during the crash. Durant would later recall in his memoir *In the Company of Heroes: The Personal Story Behind Black Hawk Down* that two SFOD-D "Delta" operators, Shughart and Gordon arrived within minutes of the crash.

In truth, Durant had been knocked unconscious during the crash. Gordon and Shughart, both snipers, were in Blackhawk Super 62 (Super Six Two) providing over-watch and cover fire throughout the course of the operation. When Super 64 went down, they requested permission to land and provide security for the crash site until evacuation; however, they were denied. Shughart and Gordon began engaging hostiles encroaching upon the crash site. This, in and of itself, is an incredible feat considering that they were firing at ground targets from a moving helicopter. Gordon and Shughart would request permission to land and secure the crash site two more times before their superior's relented and gave them permission.

Upon landing, the two offloaded and set to work. When they arrived at the crash site, they extricated Durant and his fellow crew members from the wreckage. Durant had suffered a fractured spine and right femur. The two then set up a small perimeter and began engaging hostiles as they approached. Durant later wrote, "They had fought their way through a maze of paths and shanties, driving off seemingly countless Somali gunmen. They had already done more than any two men could be expected to do. They had put their own lives on the line to try to help their fellow American soldiers."[7] MSG Gary Gordon was the first of the two snipers to be killed protecting Durant, who later wrote, "He died before I even learned his name. I will never forget him."[8]

Despite losing a companion and fellow sniper, Shughart maintained his professional repose. He retrieved Gordon's submachine gun and delivered it to Durant and asked if there were more weapons inside the wreckage of the helicopter. He retrieved the weapons, handed Durant a CAR-15, and made his final call over the radio. After receiving the message that a reaction force was en route to their location, Shughart put his radio away and moved back into his defensive position. Durant stated of Shughart, "He squinted at

---

7. Michael J Durant and Steven Hartov, *In the Company of Heroes:* (New York City, NY: New American Library, 2003), 46.

8. Ibid., 47.

the radio, stuffed it into his combat harness, hefted his weapon, and moved off around the nose of the helicopter. He left without saying another word. I would never see him again."[9]

MSG Gary Gordon and SFC Randy Shughart would both receive the Congressional Medal of Honor, posthumously, for their actions that day in Mogadishu, Somalia. The two were to be the first recipients of the CMH since the war in Vietnam. It is safe to assume that neither woke up that morning expecting it to be their last. It is a certainty that neither wished for death. However, it is highly unlikely that either, had they the opportunity, would choose to do anything differently. Their motivations were not to take lives, but to save their comrades. Michael Durant was subsequently captured and held prisoner for eleven days before being released. He would continue flying in the United States Army and retired as a CW4 in 2001.

9. Durant, *In the Company of Heroes,* 48.

NOTES:

1. Chris Hedges, "War in the Gulf: The Marines; War Is Vivid in the Gun Sights of a Sniper," *The New York Times*, February 3, 1991, p. 1.

2. Anthony Swofford, *Jarhead: A Marine's Chronicle of the Gulf War and Other Battles* (New York City, NY: Scribner, 2003), 240.

3. Swofford, *Jarhead,* 241

4. General John S. Brown "Introduction," United States Forces, Somalia after Action Report and Historical Overview: The United States Army in Somalia, 1992-1994 § (2003).

5. "United States Forces, Somalia after Action Report and Historical Overview: The United States Army in Somalia, 1992-1994 § (2003).

6. Mark Bowden, *Blackhawk Down: An American War Story* (Philadelphia, PA: Philadelphia Inquirer, 1997), 18.

7. Michael J Durant and Steven Hartov, *In the Company of Heroes:* (New York City, NY: New American Library, 2003), 46.

8. Ibid., 47.

9. Durant, *In the Company of Heroes,* 48.

CHAPTER VIII

THE WAR ON TERROR

No American is unaware of the events that took place on September 11, 2001. The situation is seared into the nation's collective memory. That morning, at 08:46 hours, the north tower of the World Trade Center in New York City was struck by a Boeing 767. Eighteen minutes later, the south tower was struck by a second Boeing 767. At 09:37, a third plane, a Boeing 757, crashed into the western side of the Pentagon in Arlington, Virginia and at 10:03, United Airlines Flight 93, a second Boeing 757, crashed into a field southeast of Pittsburgh, Pennsylvania. By 10:30 that morning, the iconic Twin Towers of New York City would be nothing more than a massive pile of rubble. In total, 2,977 innocent lives were lost.

The United States had just suffered the most devastating attack in its history. It was later determined that nineteen Islamic extremist operatives of *al-Qaeda* had hijacked the planes and carried out the plot. There was, at first, a large amount of confusion while the U.S. was searching for those responsible. It was determined that the attack was orchestrated by a terrorist organization that was part of—not one nation—but was spread out among multiple nations within the Middle East. On September 20, 2001, President George W. Bush announced to a joint session of Congress that, "From this day forward, any nation that continues to harbor or support terrorism will be regarded by the United States as a hostile regime."[1]

By the following month, the United States had identified Osama Bin Laden and *al-Qaeda* as those responsible. The President and the U.S. government insisted that the Taliban, an Islamic movement that comprised the government in Afghanistan, "immediately hand over the terrorists and close the training camps or face an attack from the United States."[2] The Taliban refused and on October 7, 2001, Operation Enduring Freedom was launched.

Soon, U.S. troops at the head of a coalition of allied nations would be operating within the Hindu Kush Mountains, searching for a non-uniformed enemy capable of blending in with the general population or disappearing into the massive cave complexes of northeast Afghanistan. By March of 2002, the U.S. Central Intelligence Agency had determined that a number of these *al-Qaeda* and Taliban operatives were moving towards the Shah-i-Kot (or Shah-i-Khot) Valley that lies southeast of Kabul, near the Afghan border with Pakistan. The 101st Airborne Division and the 10th Mountain Division of the U.S. Army in addition to U.S. Special Forces along with a coalition of allied nations and Afghan forces prepared a "hammer and anvil" attack of the valley to wipe out the enemy combatants that had amassed there.

The attack, known as Operation Anaconda, relied on American and coalition forces to be the "anvil" while Afghan forces with a contingent of Special Forces operators pushed through the valley as the "hammer." It was initially assumed that there were between 100 and 1,000 fighters in the villages of the valley; however, it was later estimated that there were only 200 – 300 lightly armed combatants.[3] Accordingly, it was determined that 400 Afghan troops, 600 U.S. troops, and 200 coalition troops would be enough to complete the mission.[4] In reality, what was encountered was between 700 – 1,000 enemy fighters

---

1. George W Bush, "Address to a Joint Session of Congress and the American People," Address to a Joint Session of Congress and the American People § (2001).

2. "9/11 Faqs," 9/11 FAQs | National September 11 Memorial & Museum, accessed September 6, 2022.

3. Richard L. Kugler, *Operation Anaconda in Afghanistan: A Case Study of Adaptation in Battle* (Washington DC: National Defense University, Center for Technology and National Security Policy, 2007), 6.

4. Ibid., 10

armed with "heavy machine guns, rocket-propelled grenades (RPGs), mortars, and even a few artillery pieces."[5]

The attack, which called for the 101st Airborne and 10th Mountain to maintain blocking positions within the valley, Objective Remington, while Special Forces occupied the eastern ridgeline was predicated on Afghan forces entering from the northwest and southwest, pushing enemy forces towards the blocking positions. In the end, the operation was considered a success. Nevertheless, the plan of attack did not survive first contact with the enemy. Afghan forces quickly found themselves outmatched and instead of attacking the interior of the valley, they were forced into a retreat. This left the 600 U.S. troops and their 200-man contingency of allied nations fending for themselves in the valley. Among them were a number of U.S. and coalition snipers, occupying the eastern ridgeline of the valley. One of these snipers, Corporal Rob Furlong of the Canadian Army, engaged and killed an enemy carrying a heavy machine gun from a distance of 2,430 meters (2,657 yards) or 1.51 miles. It was officially the longest kill by a sniper in history.

Furlong would maintain that record until 2009, when a British sniper, Corporal of Horse Craig Harrison, engaged and killed two enemy combatants at 2,475 meters (2,706 yards) or 1.54 miles. The record would again be broken in 2017 by another Canadian sniper, whose identity remains anonymous, with a kill at 3,540 meters (3,870 yards) or 2.2 miles.

Today, it is difficult for any patriotic American to imagine U.S. military snipers being considered murderers. Subsequent to Hathcock's interview, society slowly began to recognize snipers not as murderers, but as professional, elite soldiers. As First-Person Shooter video games became popular at the end of the twentieth century, the role of the sniper became coveted. This paradigm shift paved the way for snipers to feel more comfortable in detailing their exploits. Many were in the form of autobiography. As a result, legendary snipers like Chris Kyle were able to effectively share their accomplishments via the written word.

Furthermore, snipers became celebrated for the role they played in saving lives, in addition to kill counts and/or long-distance target engagements. One such story belongs to U.S. Army Ranger, First-Sergeant James Gilliland who was operating in Ramadi, Iraq in 2005, leading an echelon of snipers known as Shadow Team. While occupying an elevated hide sight known as a "crow's nest," Gilliland witnessed a fellow soldier get

---

5. Ibid., 6

shot by an enemy sniper. Gilliland immediately began scanning the area and located the enemy sniper. Unfortunately, the target was beyond the effective range of Gilliland's M-24 Sniper Weapon System. The M-24 had a listed effective range of approximately 800 yards. Gilliland ranged the enemy sniper at 1,367 yards (1,250 meters). Gilliland immediately maxed out the elevation turret on his scope and took aim at the enemy sniper. His intention was to suppress the enemy sniper to prevent him from firing upon any more friendly troops.

With his rifle D.O.P.E. (Data On Previous Engagements) set for 1,000 yards, Gilliland held another three mil-dots above the target's head. He then estimated the speed and direction of the wind and held his reticle approximately seven minutes left of the target, which translated to approximately two and a half mil-dots in Gilliland's Leupold scope. Gilliland fired, and the enemy sniper fell. Gilliland, with one shot, had eliminated the threat, saved an untold number of lives, and entered the history books for the longest kill with a 7.62 x 51 mm cartridge.[6] This is, of course, celebrated as an incredible feat of marksmanship and has even been featured in a number of television documentary programs.

However, it is also an indication that Gilliland and his team did not have the necessary equipment to effectively accomplish the mission. Had it not been for Gilliland's own training and experience supplementing the disparity of the weapon system's effective capabilities, the event may have never occurred, and the enemy sniper could have gone on to kill many more U.S. troops. The truth is that while U.S. military snipers are better equipped and armed than insurgent guerrilla fighters in the Middle East, that may not be the case when fighting a formally trained and equipped foreign military. Jared Keller, who published an article in *Task & Purpose,* cites a 2016 report commissioned by the U.S. Army that suggests U.S. military snipers have fallen behind Russia insofar as sniper equipment and training is concerned.[7]

Work on *this text.* began in 2020 when Russia was engaged in little more than saber rattling; however, in the time since, a full-fledged Russian invasion of Ukraine has taken place. Although it was not initially planned, the ongoing conflict merits its own investigation, which will be covered in a separate chapter. Another journalist, Christopher Castelli, published an article in *Inside the Pentagon* which claims that the U.S. Marine Corps is

---

6. B. James Johncox, Author's Interview with James Gilliland, October 20, 2020.

7. Jared Keller, "The US Military Is Losing the Sniper War Against Russia," Military.com, July 9, 2020.

lacking in fielding Scout Snipers and necessary equipment.[8] It does not require a stretch of the imagination to conclude that this is a direct result of the mistakes made abandoning U.S. military sniper programs in times of peace throughout American history.

Perhaps another reason for which snipers of the not-too-distant past have been demonized comes as a result of the manner by which their success is measured. Both metrics, though mentioned casually, have been referenced in this text. There are two distinct means by which snipers are considered successful; both are rather draconian. The first and most prevalent is the kill count. The kill count is often referred to by the term "confirmed kills." Confirmation of a kill is not the goal of the professional sniper. While snipers train to utilize lethal force, extinguishing a life is not necessarily the desire; rather it is the elimination of the threat that is desired. Snipers are not, however, trained to shoot to wound or maim. They are trained to engage targets in a manner which ensures the highest probability of a hit. This is known as quartering the target.

If from a long distance, the sniper can see the entire silhouette of an enemy, he/she will aim for a center mass shot, utilizing the reticle to quarter the target into four equal parts, creating the highest mathematical probability for a hit. This becomes more difficult when utilizing holds for wind or moving targets, though the principle is essentially the same. If the sniper, at closer range, can only see the head of a target, he/she will utilize the same principle. Once again, the desired outcome is a hit that eliminates the threat, which is entirely different than desiring a kill. For instance, should a sniper engage and hit an enemy who is shouldering a rifle, and the enemy suffers a severe wound, but does not die, the sniper may or may not care.

As long as the engagement prevented the enemy from taking friendly lives, the engagement is considered a success. Nevertheless, utilizing lethal force and aiming center mass more often than not results in the death of the enemy target. In any event, subsequent to an engagement, the sniper must document the event. If utilizing a spotter, working in a team environment, or providing overwatch for a ground unit moving through the area, these engagements will have witnesses who will corroborate or refute the engagement. This corroboration leads to the term "confirmed kill" as there are witnesses who can confirm the event in the documentation. Thus, as a sniper engages and eliminates threats, documentation piles up, and the sniper is credited with some number of confirmed kills.

---

8. Christopher J. Castelli, "Marine Corps Review Cites Shortage of Military Snipers in Iraq," *Inside the Pentagon* 23, no. 34 (August 23, 2007): pp. 3-4.

While there are certainly exceptions to the rule, most snipers are aware of their number of successful engagements but do not justify their existence by it.

They instead take solace in the idea that, by their actions, lives have been saved. Nevertheless, there is no accurate way to measure how many lives have been saved by eliminating one threat; therefore, the kill count becomes a standard measure of success. Perhaps there is no better evidence to support this than the story of Chris Kyle. As a result of his book, *American Sniper*, Chris Kyle received great notoriety. He is considered the deadliest sniper in American history with over 150 confirmed kills. However, outside of his mention in this book, almost no one knows that Adelbert Waldron, with 109 confirmed kills, held that record prior to Chris. Fewer still, outside of snipers, know who Simo Häyhä is. Simo, who is known as "the White Death," is considered the deadliest sniper in history for his more than 500 kills during the 100-day Winter War between Finland and the Soviet Union between 1939-40.

Furthermore, Chris himself displayed his abhorrence of being considered nothing more than a killer in a 2011 interview with Belinda Luscombe, an editor at large with *TIME* magazine. In the interview, Luscombe poses the question, "What if killing people turns out to be the thing you were better at than anything?"[9] Chris responds by saying, "Uh no, that's not true. I'm a better husband and father than I was a killer... I've got a job now I'm pretty good at. I'm pretty comfortable with not having to kill anyone."[10] Chris' response to this question is immediate and spontaneous, both indications of genuine honesty. At the end of his response, he scoffs, making known his offense at the generalization.

While Chris was considered heroic for the number of enemies he eliminated, the other measurement for a sniper's success is the long shot. The aforementioned snipers in this chapter, James Gilliland and Rob Furlong, are not widely recognized for the number of enemy targets they engaged, but rather the distance at which they engaged them. The number of variables that accompany a long-distance shot are nearly overwhelming. The most difficult to account for is wind-shift. Being that acceleration due to gravity is a constant 9.8 m/s^2, if the weight and velocity of a projectile is known, the distance that projectile will travel relative to its angle of departure is mathematically predictable. Wind-shift, on the other hand, requires a great deal more than arithmetic to conquer.

---

9. Belinda Luscombe, "Chris Kyle: American Sniper | 10 Questions | TIME," Other. *TIME*, 2011.

10. Ibid.

It requires experience. That is not to say that mathematical formulas do not exist to account for windspeed and direction; however, in the time it takes to solve the formula, the wind may have shifted, changed in speed or direction, or died out altogether. Furthermore, the wind felt by the sniper may or may not correlate to the wind at the target and in between. Many variables affect what the wind is doing, including the surrounding landscape, which can cause "switchwinds," wherein the wind speed and direction can be constant at the shooter or the target, but change dramatically somewhere in between.

In addition to the difficulty created by wind, in the case of extremely long-range shots, snipers must consider spin-drift, a shift in the path of the projectile generated by the torque of the spinning bullet as well as, the Coriolis effect, which manifests as a shift in flight path when the projectile is travelling a long distance over the curvature of the Earth. While ballistic calculators can factor in all of these advanced mathematic concepts, snipers do not always have the opportunity to leave the scope to reference a calculator. In fact, although the spotter is generally assigned the task of conducting the arithmetic and providing firing solutions, sometimes the sniper must act so quickly that the spotter can only react to the sent shot, giving corrections after the fact.

All of this is compounded by the inherent psychological stressors of combat. Thus, a sniper successfully prosecuting an extremely long-distance shot is not only an indication of a successful sniper, but an extremely experienced and gifted sniper as well. Nevertheless, the celebration of such a shot is not predicated on the extinguishing of a human life, though that may be the outcome. Rather, triumph is found in the potential that the sniper has contributed to protecting the lives of others and conquered the seemingly impossible nature of doing so.

NOTES:

1. George W Bush, "Address to a Joint Session of Congress and the American People," Address to a Joint Session of Congress and the American People § (2001).

2. "9/11 Faqs," 9/11 FAQs | National September 11 Memorial & Museum, accessed September 6, 2022.

3. Richard L. Kugler, *Operation Anaconda in Afghanistan: A Case Study of Adaptation in Battle* (Washington DC: National Defense University, Center for Technology and National Security Policy, 2007), 6.

4. Ibid., 10

5. Ibid., 6

6. B. James Johncox, Author's Interview with James Gilliland, October 20, 2020.

7. Jared Keller, "The US Military Is Losing the Sniper War Against Russia," Military.com, July 9, 2020.

8. Christopher J. Castelli, "Marine Corps Review Cites Shortage of Military Snipers in Iraq," *Inside the Pentagon* 23, no. 34 (August 23, 2007): pp. 3-4.

9. Belinda Luscombe, "Chris Kyle: American Sniper | 10 Questions | TIME," Other. *TIME*, 2011.

10. Ibid.

CHAPTER IX

2022 RUSSO-UKRANIAN WAR

It may seem counterintuitive to include a chapter focused on a war between two foreign nations in an American history text; however, as the world moves towards globalization, it becomes apparent that conflict involving a global superpower will inevitably have an impact beyond its borders that will shift American social and economic interests. Militarily speaking, it would be nothing less than dereliction to ignore Russia's ability to make war. Provided that this text is hyper focused on the minutiae of snipers and sniper-craft, this can be considered a meager contribution to that cause.

As I write this, the world is somewhat shocked at Ukraine's ability to withstand the Russian military machine. However, they are not doing so via superior military might or more technological innovation, but rather by grit and charity. Regardless of one's political view of the situation, namely the United States' involvement, it cannot be denied that Ukraine is mounting an effective defense and Ukrainian snipers, whether formally trained or by mere happenstance, are part of that defense.

In an August 2022 interview with the Associated Press, one Ukrainian sniper, who identified himself only as Andriy, provided a small but powerful account of his day-to-day duties. Going by his "war name" of "Samurai," Andriy explained that much of his work revolves around locating enemy "military positions for artillery targeting."[1] This type of reconnaissance work is typical of sniper craft, a fact well known by seasoned snipers within the trade. What makes Andriy's account particularly remarkable is that Andriy is not a sniper by trade. That is to say, his position was born of necessity. Andriy is a formally educated engineer, having left Kyiv, Ukraine to work elsewhere in Europe. Upon Russia's invasion in February of 2022, Andriy purchased an American-made precision rifle and optic and returned to Kyiv.

It is safe to presume that Andriy is a patriotic Ukrainian; however, he did not make mention of as much during the course of his interview. His motivations, instead, revolved around protecting his family who still live within the embattled city. Perhaps what was most extraordinary about Andriy's interview was his final remark. It is a sentiment shared by most, if not all professional snipers, regardless of nationality, and something that Andriy learned quite organically. He simply stated, "I don't know how to explain this: I don't like to kill people. It's not something you want to do, but it's something you have to do."[2]

Though the above account is anecdotal, it illustrates the United States' involvement, at least logistically at this point. At present, the U.S. is supplying Ukraine with the lion's share of its armaments, to include but not be limited to: FGM-148 Javelin Anti-Tank Weapon Systems, High Mobility Artillery Rocket Systems (HIMARS,) Tube-Launched – Optically-Tracked – Wire-Guided (TOW) missiles, 155mm Howitzers, 105mm Howitzers, 120mm Mortar Systems, National Advanced Surface-to-Air Missile Systems (NASAMS,) Drones, Mi-17 Helicopters, Stinger Anti-Aircraft Missile Systems, and the list goes on. Insofar as sniper rifles are concerned, the United States, in 2017, provided Ukraine with a large number of surplus M-24 Sniper Weapon Systems.[3] However, as evidenced by Andriy's account, snipers are also utilizing the free market to purchase their own equipment.

Ukraine, throughout the history of modern warfare, has displayed an uncanny knack for producing highly skilled snipers. Perhaps the most well-known is Lyudmila

---

1. Derek Gatopoulos and Adam Pemble, "Volunteer Sniper Embodies Ukraine's Versatile Military," AP NEWS (Associated Press, August 30, 2022.)

2. Ibid.

3. John L. Plaster, "Sniping in Ukraine," An Official Journal Of The NRA, accessed October 21, 2022.

Pavlichenko who, during the course of World War II, was credited with more than 309 kills, albeit she was fighting on behalf of the Soviet Union at that time.

What stands in stark contrast to other nations is Russian and Ukrainian acceptance of females within the combat arms community—more specifically, Russian and Ukrainian snipers. At the writing of this text, one such female Ukrainian sniper, Evgenia Emerald, who uses the moniker "Joan of Arc," is being celebrated not only as an heroic Ukrainian sniper, but was featured in an October 15, 2022 article in *The Odessa Journal,* published in Odessa, Ukraine for her wedding that took place on the front lines in the Kharkiv region.[4] This article demonstrates Ukraine's willingness, and perhaps even *eagerness,* to utilize and celebrate females satisfying the role of sniper.

An article published in *Newsweek* magazine on May 19, 2022, featured a Ukrainian sniper, Olen Bilozerska, discussing Russian tactics as they invaded. She likened her task as a sniper engaging Russian soldiers to a "safari,"[5] stating, "enemy vehicles moving in dense columns were destroyed by ambushes on roads passing through the forests."[6] She pointed out that in the early stages of the war, Russian forces were using tactics that have been common since World War I, stating that the enemy "primarily resorts to 'pressing-out by firing' tactics, using a large number of artillery pieces and a larger number of shells. The enemy's task is to 'grind' our positions and then try to occupy them."[7] The primary distinction between Olen and Evgenia is that Olen has been engaged as a sniper since 2014, when Russia first invaded Ukraine and occupied the Crimean Peninsula. Perhaps what these articles best illustrate is that, at least insofar as Ukraine is concerned, females are equally prepared and capable of fulfilling the role of sniper in military operations.

Insofar as Russia's 2022 invasion of Ukraine is concerned, little mainstream media coverage has been given to sniper warfare. Perhaps that comes as a result of the sniper's clandestine nature. Nevertheless, evidence suggests that Russian snipers have been used with alacrity in the Donbas region of Ukraine long before the "official" invasion of 2022.

---

4. "The Famous Ukrainian Sniper 'Joan of Arc' Got Married in the Front-Line in the Kharkiv Region," *The Odessa Journal*, October 15, 2022.

5. Gerrard Kaonga, "Top Ukraine Sniper Compares Taking out Russians to Going on Safari," Newsweek (Newsweek, May 20, 2022).

6. Gerrard Kaonga, "Top Ukraine Sniper Compares Taking out Russians to Going on Safari," Newsweek (Newsweek, May 20, 2022).

7. Ibid.

Subsequent to Russia's initial 2014 invasion and annexation of the Crimean Peninsula, which extends into the Black Sea, clandestine units of the Russian Federal Security Service, more widely known as the F.S.B., began operating in Eastern Ukraine in a region known as the Donbas. The Donbas region contains many cities that have received global attention as of late, to include Mariupol, Donetsk, and Luhansk. The region has a complicated political history.

Although the Donbas is officially recognized as part of Ukraine, pro-Russian separatists comprise a large portion of the population. Included within that are separatists who identify themselves as the Donetsk People's Republic and Luhansk People's Republic. In the aftermath of what is known as the Revolution of Dignity in 2014, the Donbas region declared their independence from Ukraine. This has since devolved into the Russo-Ukrainian War of 2022. Since that time, Russian and Ukrainian snipers alike have been extremely active in the region. A May 27, 2020, a statement from the Delegation of Ukraine to the Organization for Security and Cooperation in Europe (OSCE) reads:

> The Donbas region is also used by Russia for snipers' training. They have such sniper schools in the Rostov region, close to the Russia-Ukraine border. According to the "Jane's Defence Weekly", at the end of 2017 five sniper school graduates were deployed to Novoazovsk in Donbas. The Russian sniper schools usually concluded each training course with a month-long life–fire [sic] exercise in the Donbas region of Ukraine.[8]

This publication also identifies that in February of 2020, "the Russian FSB Special operations sniper group was detected in the temporary occupied part of Luhansk oblast."[9] What is most concerning about this particular publication is that it identifies that Russian snipers are becoming more skilled. That is not to say that they are more or less skilled than their Ukrainian counterparts; however, it should be noted that both Russia and Ukraine are producing snipers at a rate that far exceeds the United States military. The question that follows is, "are the snipers they are producing better... or more skilled... than U.S. snipers?" The short answer is that it does not matter. They need only be skilled enough.

---

8. "Statement by the Delegation of Ukraine at the 947th FSC Plenary Meeting on Russia's Ongoing Aggression against Ukraine and Illegal Occupation of Crimea," § (2020).

9. Ibid.

The Battle of Stalingrad demonstrates this. It did not matter if the Soviet Army was better than Hitler's Sixth Army; the Soviets essentially had more human capital.

This is not a suggestion that the United States military should relax its sniper training curriculum or standards; rather it should increase the number or frequency of its programs. After all, being considered elite is worthless if the program at-large is ineffective.

Where combat technology is concerned, there is a belief that due in large part to the fall of the Soviet Union, Russian combat implements have fallen behind. Indeed, Russia's technological innovation in combat technology falls short in comparison to the U.S. and its European allies. That, however, does not mean that Russia is not innovating. While still utilizing the iconic 1963 SVD, otherwise known as the Dragunov, Russia has developed the Chukavin SVCh. The SVCh, a gas operated, semi-automatic rifle that loosely resembles an AR platform, was introduced in 2018 by the Kalashnakov group and is chambered in the Russian 7.62 x 54mm R, .308 Winchester, and .338 Lapua Magnum. Although the 7.62 x 54 and .308 iterations of the weapon may be an upgrade to the SVD's accuracy potential, they do not necessarily provide an increase in effective range.

While the SVD's effective range is listed as having the potential to engage targets at 800 meters, a more realistic range is roughly 600 meters, and even then, results may vary. Major John Plaster (ret.) wrote of the SVD, "Two decades ago, I personally witnessed the SVD's performance when I was chief of competition at the European Military and Police Sniper Championships. Out to 300 yards, the SVD-armed competitors shot adequately. After that, their accuracy could not compare to their Western counterparts shooting bolt guns."[10] The .338 Lapua Magnum, on the other hand, is a large caliber that can increase a Russian sniper's effective range by approximately 62 percent. Combined with modern manufacturing techniques and a better understanding of ballistics, this is a matter of serious concern for Ukraine and the west at large.

In addition to the SVCh, the Russian company Orsis has developed the T-5000. According to the Orsis company website, this particular rifle can be chambered in .260 Remington, 6.5 x 47 Lapua, .308 Winchester, .300 Winchester Magnum, .338 Lapua Magnum, and .375 H&H. The Orsis is comparable to anything the west is producing today. A report generated by the United States Army's Asymmetric Warfare Group in 2016 stated that, "During the rapid modernization of the Russian army after 2008, the Russian army made large purchases of western made sniper rifles to include the Barret and

---

10. John L. Plaster, "Sniping in Ukraine."

Arctic Warfare Magnum (AWM). The Russian company ORSIS also makes the T-5000, one of the most capable bolt action sniper rifles in the world. These are currently the signature weapons used by Russian snipers."[11]

Ukraine is not letting this innovation go unanswered. In addition to the surplus American sniper weapon systems delivered to Ukraine in 2017, they have developed a number of precision rifle systems. One such weapon is the Zbroyar UAR-10. The system is identical in appearance to the American Knight's Armament AR-10 and fires the venerated .308 Winchester (7.62 x 51mm.) Insofar as bolt-action rifles are concerned, Zbroyar introduced the Z-008. Much like the Orsis T-5000, it is an advanced design that is comparable to anything being manufactured throughout the west. It is chambered in .338 Lapua Magnum, giving it the same 1,600 meter effective range as the .338 T-5000. It does not take much research to determine that Ukrainian snipers are taking full advantage of these rifles. Retired U.S. Army General and former CIA director David Petraeus, in a March 20, 2022, interview on CNN said of Ukrainian snipers in particular, "The Ukrainians have very, very good snipers, and they've been picking [Russian officers] off left and right."[12] Indeed, sources within Russia and Ukraine confirmed that a Ukrainian sniper engaged and killed Major General Andrei Sukhovetsky in February of 2022.

Despite these innovations, the SVD is still a capable platform when utilized as an intermediate designated marksman platform. Long range engagements make a sniper more potent, allowing them to stay outside of any effective return fire. However, urban warfare often precludes the possibility of extended range engagements. In these environments, it is the sniper's mastery of fieldcraft, his/her ability to construct effective hides that increases their survivability. This often results in shorter range engagements, inside of 300 meters. At this range, the SVD performs adequately enough to accomplish the mission, and there is no shortage of SVDs throughout Russia and Ukraine. This reinforces the argument that a sniper need not be the most elite, but they must merely be skilled *enough*. The ad-hoc sniper schools erected on the frontlines of past wars is a clear demonstration of this.

The American snipers being produced by the sniper school near Da Nang in Vietnam paled in comparison to the snipers being produced by the U.S. military today. Nevertheless, no one can make the argument that those snipers were not effective. Insofar as the

---

11. Asymmetric Warfare Group, "Russian New Generation Warfare Handbook," § (2016).

12. David Petraeus, (CNN, March 20, 2022).

Russo-Ukrainian War is concerned, if it maintains its current trajectory, that is to say, as long as nuclear weapons are not put into play, it is highly likely that Ukraine will continue to mount an effective defense.

On Wednesday, September 21, 2022, Russian President Vladimir Putin initiated a draft to raise 300,000 troops for the front lines in Ukraine. Putin insisted that he was merely activating reservists; however, a September 22, 2022, report in the *New York Times* indicated that protests erupted across Russia in response to what was obviously a conscription, whether those drafted were reservists or not.[13]

Ukraine, on the other hand, has not had to rely on such efforts as the Ukrainian citizenry has demonstrated that they are willing and able to pick up arms and fight. Thus, it is reasonable to assume that the vast majority of the snipers that Ukraine will produce throughout the course of this particular war will be trained in a manner that resembles U.S. snipers trained near Da Nang rather than those being trained by the U.S. military today. Furthermore, the crucible of combat and high stakes at play will prove equally, if not more, effective at producing operational snipers.

---

13. Anton Troianovski et al., "Ukraine War Comes Home to Russians as Putin Imposes Draft," The New York Times (The New York Times, September 22, 2022).

NOTES:

1. Derek Gatopoulos and Adam Pemble, "Volunteer Sniper Embodies Ukraine's Versatile Military," AP NEWS (Associated Press, August 30, 2022.)

2. Ibid.

3. John L. Plaster, "Sniping in Ukraine," An Official Journal Of The NRA, accessed October 21, 2022.

4. "The Famous Ukrainian Sniper 'Joan of Arc' Got Married in the Front-Line in the Kharkiv Region," *The Odessa Journal*, October 15, 2022.

5. Gerrard Kaonga, "Top Ukraine Sniper Compares Taking out Russians to Going on Safari," Newsweek (Newsweek, May 20, 2022).

6. Ibid.

7. Ibid.

8. "Statement by the Delegation of Ukraine at the 947th FSC Plenary Meeting on Russia's Ongoing Aggression against Ukraine and Illegal Occupation of Crimea," § (2020).

9. "Statement by the Delegation of Ukraine at the 947th FSC Plenary Meeting on Russia's Ongoing Aggression against Ukraine and Illegal Occupation of Crimea,"

10. John L. Plaster, "Sniping in Ukraine."

11. Asymmetric Warfare Group, "Russian New Generation Warfare Handbook," § (2016).

12. David Petraeus, (CNN, March 20, 2022).

13. Anton Troianovski et al., "Ukraine War Comes Home to Russians as Putin Imposes Draft," The New York Times (The New York Times, September 22, 2022).

PART V:

A Continuation of Politics by Other Means

"The society that separates its scholars from its warriors will have its thinking done by cowards and its fighting by fools."

-Thucydides

CHAPTER X

VIOLENCE, AN IMPLIED CONJUNCTION

Many who have come of age in the United States, and particularly those who attended public school at the end of the twentieth century and at the beginning of the twenty-first century, are quite familiar with the saying, "Violence doesn't solve anything." It is a saying typically proffered in the aftermath of a schoolyard fight, most often as the school administration is tendering punishment for said fight. However, it does not take much critical thinking to determine that the saying is a fallacy of logic. Perhaps the fight took place because one student was trying to put a stop to continued bullying, a measure of self-defense, so to speak. Perhaps the result of the fight is an end to the pervasive torture. Perhaps not. Entire texts have sought to investigate the psychology behind violence. This book is not one. However, we must accept that assigning the attributes of evil to the act of violence is unwise. Violence, like money, is amoral. It can be used for good or for evil. Adolf Hitler utilized violence to imprison Jews in concentration camps. Nevertheless, the United States, the United Kingdom, and the Soviet Union utilized violence to stop Hitler. Even then, it could be argued that the Soviet Union was only "good" insofar as the outcome of World War II is concerned. That is to say, only its use of violence was morally right. After all, Stalin's brand of communism is responsible for far more death and suffering than Hitler's concentration camps.

There are myriad examples throughout history of good conquering evil, most often via violence. Still, there are those who would argue that "might makes right" essentially claiming that "to the victor goes the spoils," to include writing the history books. That is not entirely untrue, but once again, the actual evidence must be considered on a case-by-case basis whereby objective and universal truth is carefully weighed against means and outcome. After all, one can love the United States and abhor its involvement with chattel slavery.

At any rate, violence is part and parcel of the human condition. Clausewitz believed this to be a universal truth, essentially stating that violence is a naturally occurring phenomenon while peace is a man-made construct. This is verifiably true. Since the beginning of the twenty-first century, legislated law exists to maintain peace. Otherwise, the strong would prey upon the weak without fear of consequence.

Law, or rather its enforcement, is an extension of violence aimed directly at those who would consider taking advantage of the weak or the unaware. Even something as simple as a traffic citation has the implied conjunction "or else" attached to it. That implied conjunction is an indication of a domino effect. If the citation be a simple parking ticket, the offender must pay the fine for the infraction, *or else*, be subject to confinement in a jail cell, and on it goes.

If one fails to pay the citation, and subsequently resists arrest, violence will then be utilized to subdue them. If the offender then escalates the violence during the act of resisting, higher tiers of violence up to and including the use of deadly force will be exacted. Obviously, this is an extreme example, but the principle is sound.

Insofar as the use of violence is concerned, a concept has long been utilized by modern military and police agencies. That is *speed*, surprise, and violence of action. At first glance, this seems barbarous, but this comes only as a result of the persistent attribution of evil characteristics to amoral violence. When considered critically, the concept is actually an exercise in restraint. That is to say that it does not focus on the use of excessive violence; rather the focus is the judicious application of enough discriminate violence to accomplish a mission. More simply stated, it means to strike fast and hard, so that additional force is unnecessary. Although that may explain an ideological perspective, it may be too little too late.

There is a new threat, albeit subtle, to sniper programs within the United States. It is not a threat directly to the programs but to the military and paramilitary complex at large. Although something similar surfaced in the form of the "hippie" movement of the 1960s, a new counterculture movement within the United States and other countries developed.

Some might argue that this threat has developed directly as a result of the "violence doesn't solve anything" platitude—an unintended consequence, as it were.

As of the writing of this text, multiple organizations share a common denominator. It is an abstract ideology that, simply stated, targets any institution that, at any particular point in time, may clash with their collective organization's viewpoint. More recently, this concept has been termed "presentism" and has been discussed at length by historians. Indeed, it has become a hotly debated topic with advocates on both sides extolling or condemning its principles. Presentism is assigning present day virtue to people or events of the past without critical consideration, or more simply stated, judging past events through the lens of modern morality.

It is a difficult balance to maintain for an historian. After all, slavery and human trafficking are disgusting practices, but for one to say that there is no way they would have been involved in said practices should they have lived during that era is highly disingenuous. That is not to say they do not *believe* they would have. It is a certainty that their moral and ethical convictions guide their actions in the present. However, the presumption that an individual would have stood in obstinance to what was then considered socially acceptable is dubious at best. Where this begins to pose a threat to history is when self-termed "social justice warriors" seek to strike from the history books any who did not then display what is considered moral today. Thomas Jefferson, author of the Declaration of Independence and the third president of the United States of America is one example. Jefferson was both a founding father of the United States and a slave holder. Present day morality dictates that being a founding father is *good* and participating in slavery is *bad*. Most agree that slavery was a terrible chapter of United States history, but due to the detrimental effect of polarization and identity politics, it has become difficult to approach that particular subject.

Thus, the concept can be approached from a different direction, the writing of the Declaration of Independence. Had the patriots been defeated during the Revolutionary War, it is entirely possible that present day Americans would still identify as British subjects, which in turn may have meant that Thomas Jefferson would be considered a traitor and Benedict Arnold a hero. Lynn Hunt, a professor of Modern European History at UCLA, wrote of the ideology:

> "Presentism, at its worst, encourages a kind of moral complacency and self-congratulation. Interpreting the past in terms of present concerns

usually leads us to find ourselves morally superior; the Greeks had slavery, even David Hume was a racist, and European women endorsed imperial ventures. Our forbears constantly fail to measure up to our present-day standards."[1]

Much like the past, these ideologies lead to the creation of popular organizations. Some of these organizations, such as "ANTIFA" (Anti-Fascist movement) and BLM (Black Lives Matter,) will often utilize demonstrations, violent and non-violent, to protest the status quo, as well as, institutions that are predicated on the use of discriminate violence, all the while claiming moral superiority. This was evident during the riots that took place throughout American cities in the summer of 2020. On multiple occasions, police departments, for instance, were targeted and burned to the ground whether or not the police department being targeted had been involved in a use of force incident. These organization's ethos are malleable, changing to suit whatever particular outcome is desired at any particular time and are often only loosely affiliated with other member organizations—usually in name only.

That being said, these organizations are making enough noise to garner national attention, whether it be in the form of media coverage or political punditry. When the attention does not favor their cause, they have historically relied on decrying their opponents as "racists" as a means to shut down the conversation. Whether the argument concerned race is of no concern. While they may seem altruistic on the exterior, a closer look reveals that their true motivation is to obtain power via the aforementioned attention, which means that no topic is "off limits" as long as their conclusions contribute to their cause. Sometimes these conclusions require an almost impressive display of verbal gymnastics to tie one justification to the next.

It may seem as though it has nothing to do with snipers; however, what must be considered is the various tenets of the ideology, one of which maintains that by virtue of being weak, one is "good," and one who is formidable must, therefore, be "bad." In a 2018 interview with John Stossel, Dr. Jordan Peterson, a clinical psychologist, author, and Professor Emeritus at the University of Toronto, stated that "Not being violent is not a virtue. The capacity for danger and the capacity for control is what brings about

---

1. Lynn Hunt, "Against Presentism," *Perspectives on History,* The American Historical Association, May 1, 2002.

the virtue. Otherwise, you confuse weakness with moral virtue."[2] This ideology of virtue via weakness has become pervasive, influencing not only young men, but experienced leaders as well. That is evidenced by the fact that the highest-ranking leaders in the United States military have included some of these organization's ideological texts in their recommended readings. Furthermore, U.S. military branches have invested extensive capital in recruitment ads focused on individual identity as opposed to patriotic or teamwork themes. Make no mistake, while military and paramilitary organizations such as law enforcement exist to provide safety, it is accomplished via the ability and willingness to do violence, if—and only if—the need should arise.

To suggest otherwise is nothing short of malfeasance. Thus, individual identity within military and paramilitary organizations is near the bottom on the list of priorities, whereas learning to work as a team, as well as do violence on the behalf of others that cannot or will not is at the top. That said, the new recruitment initiatives being displayed only serve to weaken a military predicated on teamwork, and by extension the various team-centric programs within the military, to include snipers.

While there are a number of reasons these counterculture movements materialize, they are all similar insofar as their inability to critically analyze long term impacts. Typically, they coalesce around what is, and should be, a positive message. For instance, the Black Lives Matter movement maintains that the lives of black and other dark-skinned people are invaluable. This is absolutely correct, and few would disagree; however, should one point out that they are no more and no less valuable than the lives of other decent human beings, the fallout is catastrophic.

Thus, the only way to maintain good standing within the BLM movement, for example, is to claim or agree that black lives, to the exclusion of all other lives, matter. Antifa's title, likewise, is merely a shallow platitude. Fascism is indeed a terrible ideology; however, members of the Antifa counterculture movement do not seem to understand that their own behavior more resembles fascism at work than does those whom they protest. The long-term impact is that, inevitably, some of those who have powerful platforms and the ability to reach the masses will take up and parrot the rhetoric of the cause, influencing those who are most susceptible to ideologies that are based in emotion rather than logic, which are typically teenagers and young adults. This results in creating division that manifests as tribalism within a community. The problem is not that these messages or

---

2. John Stossel, interview with Dr. Jordan B. Peterson "Jordan Peterson, the FULL interview," June 2018.

organizations exist. After all, the first amendment to the U.S. Constitution protects and guarantees their right to exist. A thought experiment that might be conducted is to visualize a utopian society wherein the concept of "bad" or "evil" no longer exists. Without it, there is no longer a contrast for that which is "good" or "righteous." Thus, eliminating the "bad" thoughts and ideologies, or even the ideologies of those with whom one disagrees is not a viable option.

The problem is, instead, the incentivization of objectively bad ideas over objectively good ideas. Albert Einstein famously stated, "what is right is not always popular and what is popular is not always right." A critical thinker might ask, "What was the inspiration behind this quote?" Unfortunately, bad ideas are sometimes accompanied by immediate gratification; thus the bad idea has high incentive that manifests in a number of ways, to include popularity; whereas, the good idea's incentives, such as a favorable economy, equitable opportunity, or political stability, take much longer to realize. Naturally, the question that follows is, "How does one convince an entire society that immediate gratification is not preferable to long-term stability?" He or she that manages to answer this question without infringing upon constitutional rights will be rich indeed.

Until then, these groups contribute, albeit sometimes unintentionally, to the idea that those within the warrior class, and especially snipers, are somehow violent sociopaths. It is for this reason that, in years past, they have been labeled "Murder, Inc." Contrary to what the Murder, Inc. epithet suggests, snipers are not murderers. Nevertheless, the pejorative has been applied over the years. This epithet was not originally coined to describe snipers. In fact, it was a moniker for a squad of hitmen that worked for the National Crime Commission established by several mob bosses in the 1930s. This might seem like an unimportant detail, but the equivocation of snipers to gangland murderers is far more disturbing than one might initially think.

With this in mind a portion of this chapter is dedicated to distinguishing the difference between a murderer and a professional sniper. It is an all-too-common occurrence that one turns on the television to watch the local and/or national news and is confronted by a grisly murder scene. Sometimes, it is accompanied by images of deceased bodies littering the ground. In the event that the killer used a rifle . . . any kind of rifle, but particularly those with some kind of optic . . . the murderer is identified as a sniper. Most listen and watch, some in horror . . . others somewhat apathetically . . . and the voice of the news anchor continues while subconsciously the thought, "Thank God that didn't happen here" plays in the back of the mind. Investigations are launched; hearings are held;

reporters dig; and the world continues to turn. No one seems to notice that the word "sniper" was once again tied to a horrible act.

There is perhaps no better example of this than the "Beltway Snipers" a.k.a. the "DC Snipers" incident that took place during the month of October 2002, in Washington DC, Maryland, and Virginia. The investigation began when James Martin, who was 55 years old, was gunned down at approximately 18:30 hours on October 2 in the parking lot of a Shopper's Food Warehouse, a grocery store in Glenmont, Maryland. The killers, John Allen Muhammed, who was forty-one years of age, and Lee Boyd Malvo, who was only seventeen, would go on to kill nine more and critically injure three before it was over. Subsequent to their arrest, Malvo confessed that the pair had committed far more murders than just those in the District of Columbia metropolitan area. It was determined that the two were responsible for murders and/or robberies in Alabama, Arizona, Florida, Georgia, Louisiana, and Texas in addition to those in the Washington DC area. In total, seventeen people had been killed and ten injured. When captured, the two were in possession of a Bushmaster AR-15 rifle chambered in .223 caliber that featured a bipod and a holographic optic. The FBI's "Famous Cases" website incorrectly identified the bipod as a tripod, which is concerning considering that the FBI is considered the U.S.'s preeminent law enforcement agency. It also refers to the holographic optic as a "rifle scope for taking aim."[3]

While the latter is not technically incorrect, the presumption could be made that the vernacular utilized is far more dramatic than "holographic gun sight." This may seem to be a technicality that does not have any bearing on the circumstances. After all, the definition of the word *sniper* does not predicate its use specifically for military or law enforcement personnel. Nor does it indicate that a rifle equipped with a telescope is necessary. Indeed, Oxford's dictionary defines the word sniper as, "a person who shoots at somebody from a hidden position,"[4] and Merriam-Webster dictionary defines it as "to shoot at exposed individuals (as of an enemy's forces) from a usually concealed point of vantage." [5] Seemingly, the only requirement of a sniper is a firearm and a hiding place. Malvo and Muhammed had both.

---

3. "Beltway Snipers," FBI (Federal Bureau of Investigation, May 18, 2016).

4. "Sniper," sniper noun - Definition, pictures, pronunciation and usage notes | Oxford Advanced Learner's Dictionary at OxfordLearnersDictionaries.com (Oxford Learner's Dictionary), accessed September 5, 2022.

5. "Sniper Definition & Meaning," Merriam-Webster (Merriam-Webster), accessed September 5, 2022.

The two had made modifications to a blue 1990 Chevrolet Caprice that allowed access to the trunk through the back seat. They had fashioned a loophole above the rear bumper by cutting a small square that allowed Malvo to fire from within the trunk of the vehicle. It would appear as though Malvo and Muhammed had been correctly labeled. Despite the definition, a collective decision must be made to separate the word sniper from those who commit wanton acts of violence. A sniper is far more than what the official definition imparts. This can be explained technically, as well as ideologically.

Firstly, the original usage of the word was a description of a marksman so gifted that he could hit a swift game bird that was indigenous to India. There is no mention of a hiding place or concealment of the marksman. As technology advanced, these gifted marksmen increased the distance at which they could successfully engage targets. That is not to say that a sniper is only defined by the "long shot," but rather that a sniper is a master of his/her weapon platform, capable of engaging a target at any range that the rifle is effective. Shots within 300 meters are not difficult and can be accomplished with rudimentary training. For instance, the United States Army trains recruits to fire at this distance with iron sights during Basic Rifle Marksmanship training.

The United States Marine Corps trains its marines similarly out to 500 meters. The trained sniper is capable of delivering accurate and precise rifle fire well in excess of these ranges. Though a large degree of importance is placed on it during training, concealment or camouflaging techniques are incidental. Ideologically speaking, as previously mentioned, a sniper is one who desires to protect life. Though it be within his/her ability, the taking of lives is not desired . . . merely necessary to accomplish the primary objective.

Malvo and Muhammed may have had a hiding place; however, their motives were nefarious. The two wished to sow terror within the population. In stark contrast to the primary objective of a true sniper, the two targeted innocent individuals in order to realize some perverse machination of revenge. Neither was out to save lives. Furthermore, none of the shots that they took were at distances that could be considered extreme. Suffice to say, the two were murderers . . . period. They were captured on October 24, 2002, tried, and convicted. Malvo is currently serving six life sentences, and Muhammed was executed by lethal injection in 2009.

It may seem inane to devote a chapter to this subject. Most people can determine the difference between a real sniper and a murderer based upon the context of the discussion. Nevertheless, the attributes of something good are rarely attributed to something bad. For instance, should someone wear MARPAT camouflage while robbing a bank, it is doubtful that the headline would read "Marine Robs Bank." Furthermore, it is a slap in the face to those who have dedicated their lives to a science that is not easy by any stretch of the imagination. Finally, it is ignorance on display. That is not meant to be derisive. Truly it is a lack of knowledge and understanding. It is safe to assume that the vast majority of reporters are not snipers; thus they do not understand the implications of their writing. It is an honest mistake, but a mistake none the less.

Correcting this mistake is difficult. Of the United States' population, comparatively few volunteer for military and/or police service. Of those, even fewer become snipers. It is difficult for those few who go on to become elite marksman to compete for the title, so to speak. Professional military and/or police snipers outnumber deranged killers; however, professional military and/or police snipers, by design, seldom make headlines. The title of *sniper* is provocative and will continue to grab attention and wonder, but it belongs to those who have earned it, who train hard and dedicate themselves to saving lives . . . not to those who wantonly take them.

NOTES:

1. Lynn Hunt, "Against Presentism," *Perspectives on History,* The American Historical Association, May 1, 2002.

2. John Stossel, interview with Dr. Jordan B. Peterson "Jordan Peterson, the FULL interview," June 2018.

3. "Beltway Snipers," FBI (Federal Bureau of Investigation, May 18, 2016).

4. "Sniper," sniper noun - Definition, pictures, pronunciation and usage notes | Oxford Advanced Learner's Dictionary at OxfordLearnersDictionaries.com (Oxford Learner's Dictionary), accessed September 5, 2022.

5. "Sniper Definition & Meaning," Merriam-Webster (Merriam-Webster), accessed September 5, 2022.

CHAPTER XI

THE FUTURE OF SNIPER WARFARE

Human nature seems to require that a single action be identified as a "reason" for unfortunate circumstances, especially those that deteriorate into catastrophic events. When a murder takes place, the community wants to know *why* the murderer did it. Another example, albeit on a much larger scale, is that children (especially Americans) are taught in school that the single reason for the eruption of the first World War was the assassination of Archduke Franz Ferdinand and his wife, the Duchess Sophie of Hohenberg. This suffices for grade school lesson plans, especially since the U.S.'s involvement in World War I was limited to its final year. Nevertheless, it reinforces the idea that all events boil down to one single act, and that had the single act gone in the other direction, everything would have turned out fine. The truth is no one can be certain of what might have been. Given that Gavrilo Princep's actions that day in Sarajevo were the spark that lit the fuse, the powder keg had been in place since the defeat of France in the Franco-Prussian War of 1870-71. It is arguable that Otto Von Bismarck, who unified the various Germanic kingdoms into the single nation of Germany after France's defeat in 1871, manifested most of the circumstances required to ensure a Great War (at the very least) would take place.

Thus, the assassination of the heir apparent to the Austro-Hungarian throne was not so much the beginning of the war; rather it was the end of the peace. Nevertheless, there is always a beginning to the "butterfly effect," as it were, and it is more often than not, somewhat innocuous and almost never "glamourous."

An article *Inside the Pentagon* was written by Christopher Castelli in the midst of Operations Iraqi Freedom and Enduring Freedom in 2007. The article claimed that the U.S. Marine Corps was lacking in fielding Scout Snipers and necessary equipment.[1] Jared Keller, who published an article in the July 2020 edition of *Task & Purpose* cites a 2016 report commissioned by the U.S. Army that suggests U.S. military snipers have fallen behind Russia insofar as sniper equipment and training is concerned.[2]

Unfortunately, it seems as though neither article found its mark—at least not among its intended targets. It does not require a stretch of the imagination to conclude that both articles were intending to address problems created by abandoning U.S. military sniper programs in times of peace throughout American history.

Those who do not learn from history are doomed to repeat it . . . at least that is how the saying goes. Nevertheless, in February of 2023, the Commandant of the Marine Corps, General David Berger, and his second in command, General Eric Smith, announced an end to the vaunted Marine Scout/Sniper program in favor of reconnaissance platoons. When questioned about the shift towards the reconnaissance model, General Berger stated: "If you're so big and fat and immobile and vulnerable to their sensors, all the lethality in the world ain't going to help you."[3]

Although the response is somewhat vulgar, there is something to be said for a general who speaks plainly. Of course, at first glance . . . it would appear as though the general is speaking in terms of physical fitness . . . and that could partially be the case; however, it is more likely that he was speaking of programs that require significant echelons of support infrastructure, which in turn decreases the mobility of said programs. Indeed, since the first and second world war, Jominian wars of mobility have supplanted Clausewitzian methods of forcing decisive battles. Notwithstanding, the question remains. Is eliminating the Marine Corps' Scout/Sniper Platoons and training programs a good idea?

---

1. Christopher J. Castelli, "Marine Corps Review Cites Shortage of Military Snipers in Iraq," Inside the Pentagon 23, no. 34 (August 23, 2007): pp. 3-4, https://doi.org/https://www.jstor.org/stable/insipent.23.34.08.

2. Jared Keller, "The US Military Is Losing the Sniper War Against Russia," Military.com, July 9, 2020, https://www.military.com/daily-news/2020/07/09/us-military-losing-sniper-war-against-russia.html.

3. Megan Eckstein, "Marines' Force Design 2030 Update Refocuses on Reconnaissance," Defense News, August 19, 2022, https://www.defensenews.com/naval/2022/05/09/marines-force-design-2030-update-refocuses-on-reconnaissance/.

A July 2023 article in the *Marine Corps Times* titled "Doubts about scout snipers arose in infantry units, No. 2 Marine says" further expounds on the thinking behind this decision. In the aforementioned article, General Smith provides a number of reasons why the decision was made. Smith stated that the idea originated as a grass roots movement at the division level, with the four division commanders stating, "Hey, we need more scouting. We need ISR—intelligence, surveillance, reconnaissance. We need less precision shooting..."[4] Smith maintains that if the average Marine rifleman "didn't hit the 700-, 800-meter target" they were "chastised, you know, for being a loser." [5]

There is no doubt that ISR (Intelligence, Surveillance, and Reconnaissance) capabilities are paramount in warfare (and to a greater or lesser extent, to prevent warfare.) Indeed, the need for gathering information, surveillance, and reconnaissance in combat environments could be considered common sense. Warfare throughout recorded history has relied on no small amount of information gathering to secure victory. After all, *The Art of War* by Sun Tzu was written some 2,500 years ago, and he was quite fond of the use of spies, even going as far as to refer to them as a "ruler's greatest treasure."[6] Perhaps the best example of superb ISR capabilities were the various clans of Iga and Koga during the Warring States period of fifteenth and sixteenth century Japan.

Their use of Shinobi gave rise to what is known as a ninja today, and while the ninja is often characterized as an assassin, it was their ability to infiltrate and secure information that allowed the State of Iga to survive against the much more powerful Daimyo, Oda Nobunaga, for as long as they did. Nevertheless, neither of the above-mentioned historical references eliminated their missile range combatants to facilitate ISR gathering. They may not have been mutually exclusive, but neither were they one and the same.

Those who study past wars have the benefit of perfect sight and can clearly articulate why snipers were necessary in each conflict studied. There is, however, no way anyone can accurately and precisely predict what future battlefields have in store. Those who try to predict it seldom get it right. The lessons of the past clearly articulate that eliminating sniper programs was a poor choice in every instance. Snipers are adept at intelligence

---

4. Irene Loewenson, "Doubts about Scout Snipers Arose in Infantry Units, No. 2 Marine Says," Marine Corps Times, July 6, 2023, https://www.marinecorpstimes.com/news/your-marine-corps/2023/07/06/doubts-about-scout-snipers-arose-in-infantry-units-no-2-marine-says/.

5. Ibid.

6. Sun Tzu, The Art of War, trans. James Trapp (New York City, New York: Chartwell Books, 2012), 93.

gathering . . . but have the added benefit of dispensing lethality with surgical precision. The same could be said of a Predator ISR drone, though the collateral damage of a hellfire missile is more significant than that of a single .300 Winchester Magnum projectile, not to mention the cost/benefit ratio. At any rate, neither tool should be cast aside; each was utilized for different outcomes. Each had the right tool for the job. Furthermore, there is an added benefit in the use of snipers. A silent dread settles over a battlefield that is being surveyed by a sniper. Even if the sniper is no longer there, the fear that he *might be* has an incalculable impact on potential combatants.

Within the American law enforcement community, officers acknowledge that the very first level of force on the "Use of Force Continuum" is the officer's presence. It is an understood use of force as most will cease foolish behavior in the presence of a uniformed officer. This observable shift in behavior occurs due to the mere fear of arrest. Now imagine a soldier on a battlefield knowing that a sniper may be surveying a particular salient. The specter of spontaneous death is a powerful inhibitor. There is no possible way of knowing how many lives have been saved by this dynamic, but it is most assuredly more than the number of lives taken. Of course, the same could be said of a Predator/Reaper Drone operating in theater; however, certain human intuition must also be considered. For instance, as of 2021, an AGM-114 Hellfire II missile, to include training and technical support, costs on average between $130,000 and $160,000 per unit.[7]   For fiscal year 2021, the Pentagon requested $516.6 million for Hellfire missiles alone.[8]   At an average cost of $140,000, this equates to 3,690 missiles. An individual enemy combatant, not to be confused with an enemy HVT (high value target,) in an urban environment may believe that the U.S. is not willing to spend that much or risk collateral damage in order to eliminate him. Thus, there exists a degree of "safety" which further emboldens that individual to continue his activity. The same degree of safety does not exist when the combatant is aware that there may be a sniper in the area and that there are no cost inhibitions related to a single round that creates virtually zero collateral damage.

Insofar as General Smith's claim that Marines chastise one another for not being able to engage an 800-meter target, this may be anecdotal, or objectively true, or it may be an

---

7. "Ultimate Guide on AGM-114 Hellfire Missiles: Capabilities and Cost," The Defense Post, March 30, 2021, https://www.thedefensepost.com/2021/03/22/agm-114-hellfire-missile/.

8. Office of the Under Secretary of Defense (Comptroller)/Chief Financial Officer, Program Acquisition Cost By Weapon System: United States Department of Defense Fiscal Year 2021 Budget Request (2021), https://comptroller.defense.gov/Portals/45/Documents/defbudget/fy2021/fy2021_Weapons.pdf.

exaggeration made "off the cuff." What must be considered is that the technology required for making consistent precise and accurate long-distance shots (for the purpose of this text, those that exceed 300 meters) has existed since the middle of the nineteenth century (some 160 years) and despite this, only a certain few have demonstrated the necessary talent, discipline, temperament, and fortitude to consistently and successfully eliminate an enemy combatant at or exceeding that range.

After all, not every Marine is a sniper and not every sniper is a Marine. Nevertheless, the last eight Marine Corps Scout/Snipers for the foreseeable future graduated Scout/Sniper school on December 15, 2023.[9] Unlike their predecessors, however, they were not granted the occupational specialty identifier awarded to all Scout/Snipers who came before them. Whether or not this will be a permanent change cannot be known, though history would seem to indicate that it will not. The generals, and their subordinates within the command structure, seem to take issue with the inflated infrastructure required to support a sniper element, essentially indicating that it is a cost/manpower issue. There is no way of knowing what was said in those conversations. It is also unknown whether any of those particular officers were directly in charge of Scout/Sniper Platoons; however, considering the decision was seemingly unanimous, it is probably a safe assumption that they were not. That is not to say that the commandant and his subordinates acted in a nefarious way, but rather out of ignorance.

The truth is that a sniper training curriculum could be pared down, modernized, or custom built for a particular objective. For instance, knowing how to stalk through the jungles of southeast Asia is good training, but is it important right now? Is it a course that could be a standalone class, reserved for later training and/or combat objectives? Emphasis could be shifted towards building effective urban hides and utilizing them to gather more intelligence. A more versatile Scout/Sniper School curriculum could be drafted with greater focus being placed on the needs of the corps.

Finally, there is an implied justification for this decision. The various commanders via their collective determination have indicated that the Scout/Sniper program has resulted in some form of direct or indirect mission failure. For example, as a result of the associated cost, or the manpower necessary for the Scout/Sniper program, mission X, Y, or Z were not accomplished. Otherwise, what justification would there be to conduct a critical

9. Irene Loewenson, "The Marine Corps Has Trained Its Final 8 Scout Snipers," Marine Corps Times, December 28, 2023, https://www.marinecorpstimes.com/news/your-marine-corps/2023/12/28/the-marine-corps-has-tr ained-its-final-8-scout-snipers/.

analysis? It is unlikely that the various commanders would attribute mission failure to their own action, inaction, or failure to properly utilize a tool. A fitting analogy might be a technician, without knowing what future jobs have in store, arbitrarily deciding that because a specific tool is too heavy or too expensive, it must therefore no longer be necessary; thus the technician discards the tool, despite the fact that the tool has been utilized to great effect throughout the course of the technician's career. It is a myopic point of view. At any rate, there are two absolutes with regard to this decision. First, another war will come, and a second; snipers have been necessary in every war in which Americans have fought.

Few have paid attention to the abolition of the various sniper programs throughout American military history, and fewer still are those who have speculated as to why. As mentioned earlier in this chapter, there is a desire to continuously narrow down causation until one clear reason remains. Pegler suggests that the phenomenon was a response to the culmination of five dynamics, the first of which suggests that too few high-ranking officers in the military supported the proliferation of sniper programs, which ties into the second dynamic that suggests being a commander of a sniper unit was not conducive to a successful career in the military. He also points out that sniper equipment and training are cost prohibitive.

Both Generals Smith and Berger have, as of 2023, demonstrated how pervasive those arguments are. However, his last two premises go hand in hand. They state that sniping is a clandestine profession comprised of men who are unable or unwilling to discuss their experiences, which leads to the development of a "catch-22" with the fifth and final reason being that "there existed a distinct antipathy towards snipers as combat soldiers, with the lingering attitude that it was somehow an unfair way to wage war."[10] This last sentiment has been echoed over the centuries, especially in the American historical record. Whether it was referred to as a pejorative "American Way" of war, or later snipers being referred to as "10 cent/13 cent/ *insert cost of a cartridge here* killers," or "Murder, Inc.," the idea that a sniper is somehow a cold-hearted, unfeeling, calculated, and psychopathic killer has been pervasive since before an official title even existed for the discipline.

As far as the first three of Pegler's dynamics are concerned, while many officers may have had apprehensions about establishing and leading sniper units, *many* officers were not required, and, while the second of the arguments may be true in part, to determine

---

10. Pegler, *Sniper: A History of the US Marksman*, 149.

whether a military officer's career will be successful is incumbent upon a large number of variables that may include, but is certainly not limited to, units which they command. The cost prohibition aspect of snipers is an argument that can easily be dismantled. During the Korean War, for instance, one napalm bomb cost $40.00 to create.[11]   Each napalm bomb contained roughly 100 gallons of napalm fuel weighing approximately 73 pounds.

Every day 250,000 pounds of napalm was dropped on Korea. That is approximately 3,400 bombs per day at a cost of $136,000.00 per day. As far as ammunition, Snipers were equipped with the same Garand M1 rifle as line troops, although it had modifications. The Garand fired a Springfield 30-06 (.30 caliber, model of 1906) cartridge that was roughly $0.10 per round, meaning, if comparing straight across the board, snipers would have had to fire approximately 1,360,000 rounds, per day, to equate in price to the use of napalm.

That is not to say that the United States should have used only one or the other, but rather to illustrate that the argument that the cost of training and equipping a sniper being too great is disingenuous at best. Furthermore, the equipment utilized by a sniper, with the exception of the projectile fired, has a *slightly* longer service life than a napalm bomb. For instance, World War I era Springfield M1903 rifles are still owned and fired by collectors today and of well-kept examples, most, if not all, have lost little in terms of accuracy. In addition to this, the implementation of even one sniper establishes a force multiplication dynamic that not only is invaluable for gathering intelligence and performing counter sniper operations, but also inflicts an intangible psychological stressor on the enemy, exponentially slowing enemy operations. It could be argued that the first three are all entirely predicated on the analysis of the last two of these five considerations.

Critically thinking about the last two sniper dilemmas produces a "chicken vs. egg" argument. Did snipers receive contempt from their peers due to their clandestine nature, or did their clandestine nature evolve as a result of the contempt they received from their peers? The question is, of course, rhetorical and more than likely depends upon the individual sniper, commander, unit, etc.

What is not debatable is that, more so than any nation prior to or since, the United States, from its birth to the modern era, is inextricably tied to the marksman and the rifle. Nevertheless, insofar as martial force is concerned, no other nation has so readily aban-

11. John   Pike,   "Napalm   in   War,"   GlobalSecurity.org,   accessed   August   4, 2022, https://www.globalsecurity.org/military/systems/munitions/napalm-war.htm#:~:text=During%20the%20Korean%20War%2C%20the,each%2C%20and%20held%20100%20gallons.

doned its marksmen upon the completion of the mission. The complex answer is that it is a combination of variables, predicated upon the era that resulted in sharpshooter/sniper programs being abandoned. In early American history, the tactics conflicted with the conventional warfare doctrine of the age. By the early twentieth century, sniping was relegated to merely being a necessary product of trench warfare. By the mid twentieth century, it was considered cost prohibitive. Though these attributes shifted with the era during which American wars were fought, only one variable remained the same throughout, to which all the others contributed to the moral and/or ethical repugnance of the profession. This is confirmed by the fact that the top two commanders of the United States Marine Corps in 2023, and their subordinate division commanders, still hold some implicit bias towards snipers, relegating the discipline to mere "precision shooting."

It seems it is a denigration born of ignorance or jealousy, perhaps, rather than disgust. Nevertheless, though the offered justifications for eliminating the various sniper programs within the U.S. military may be disguised in the nuance of budget or manpower concerns, Occam's razor suggests that those are, at best, decisions born of ignorance, and at worst, willful blindness.

Thus, it is reasonable to assume that this sometimes explicit, sometimes implicit, bias against snipers has had the most damaging impact. Identifying these various problems establishes a path to follow, or rather a path to avoid. What now matters, and will continue to matter, is the state of American military preparedness. For those who subscribe to the Clausewitzian principle that suggests war is a naturally occurring phenomenon, and that peace is merely the manmade construct that occupies the space between, it is absolutely imperative to correct these mistakes of the past so that the defensive capabilities of the United States military are not further neglected or surpassed by potential enemy nations.

Therefore, the one answer, the beginning of the butterfly effect, as it were, is the contempt with which snipers are held. Through careful analysis, it seems somewhat easy to identify. Unfortunately, it is difficult to fix. It will require a substantial amount of capital for training, equipment, and manpower, and perhaps a less likely event, recognition from top tier commanders that mistakes have been made, to include recently. Furthermore, it will require a great deal of understanding from the American public at large. Considering the funeral procession for Chris Kyle was over a mile in length, recognition of one of America's finest heroes is a sign that the nation is headed in the right direction, although the speed needs to be increased.

As far as snipers themselves are concerned, it will require more of the steady resolve, iron constitution, and indomitable spirit that has been displayed thus far. It is not difficult for them. They embrace the loneliness and the absent recognition. The discipline is not, and never has been, dependent upon acknowledgement. In fact, the argument could be made that acknowledgement is the opposite of what a sniper strives for.

Naturally, the question that follows is: "What will happen to snipers in the future?" Obviously, there is no way to know for sure. One can only look to history as a blueprint. As of October 7, 2023, the terrorist organization Hamas, which is also the elected Palestinian government of the Gaza Strip, as it were, invaded Israel and murdered nearly 1,200 individuals, including babies, young children, and the elderly.[12] In addition, Hamas fighters abducted several hundred hostages, some of which are still being held at the writing of this text. Following Israel's subsequent response, to include the invasion of the Gaza Strip, Iran escalated its involvement. Whereas Iran once merely funded the various terrorist organizations, to include Hamas and Hezbollah (headquartered in Lebanon on Israel's northern border,) the Iranian government, on April 13, 2024, began launching directed ballistic missile and drone strikes at Israel.

This, in addition to the ongoing conflict in Ukraine, has set the stage for another global conflict. At present, the United States and other NATO allies are providing cash and materiel to Ukraine, a non-NATO country. The United States is also providing funding to Israel in its efforts against Iran and the various terrorist fundamentalist groups being funded by Iran. Within the United States, politicians argue merits on both sides of the aisle, some condemning the involvement of the U.S. It is not difficult to find characteristics similar to the beginnings of both previous World Wars, seemingly marinating in a vat of Cold War policy. That is to say, the powder keg has once again been placed and lacks only a spark. Those who study history might speculate that another major global conflict is more likely than not. If so, the major cause for concern will undoubtedly be the use of nuclear weapons. Even so, wars cannot be won without boots on the ground.

The Vietnam War demonstrated that the act of destroying an enemy stronghold, and subsequently leaving it unoccupied, will only result in the enemy reoccupying said stronghold. Thus, a bombing campaign, like that of Operation Rolling Thunder on North Vietnam, will likely only generate temporary results without the subsequent in-

---

12. Daniel Byman et al., "Hamas's October 7 Attack: Visualizing the Data," CSIS, accessed April 22, 2024, https://www.csis.org/analysis/Hamass-october-7-attack-visualizing-data.

fusion of troops to occupy territory. To assume that enemy combatants will not utilize snipers would not only be a mistake, but an act of dereliction. In fact, it is reasonable to assume that the technology gap that exists between the United States and its potential enemies would necessitate the use of snipers by the enemy, as a means by which the playing field could be somewhat leveled. After all, it turned the tide of battle for Americans when they fought the most powerful empire the world had ever seen.

NOTES:

1. Christopher J. Castelli, "Marine Corps Review Cites Shortage of Military Snipers in Iraq," *Inside the Pentagon* 23, no. 34 (August 23, 2007): pp. 3-4, https://doi.org/https://www.jstor.org/stable/insipent.23.34.08.

2. Jared Keller, "The U.S. Military Is Losing the Sniper War Against Russia," Military.com, July 9, 2020, https://www.military.com/daily-news/2020/07/0 9/us-military-losing-sniper-war-against-russia.html.

3. Megan Eckstein, "Marines' Force Design 2030 Update Re-focuses on Reconnaissance," Defense News, August 19, 2022, https://www.defensenews.com/naval/2022/05/09/marines-force-desig n-2030-update-refocuses-on-reconnaissance/.

4. Irene Loewenson, "Doubts about Scout Snipers Arose in In-fantry Units, No. 2 Marine Says," Marine Corps Times, July 6, 2023, https://www.marinecorpstimes.com/news/your-marine-corps/2023/0 7/06/doubts-about-scout-snipers-arose-in-infantry-units-no-2-marine-says/.

5. Ibid.

6. Sun Tzu, *The Art of War*, trans. James Trapp (New York City, New York: Chartwell Books, 2012), 93.

7. "Ultimate Guide on AGM-114 Hellfire Missiles: Capabilities and Cost," The Defense Post, March 30, 2021, https://www.thedefensepost.com/2021/03/2 2/agm-114-hellfire-missile/.

8. Office of the Under Secretary of Defense (Comptroller)/Chief Financial Officer, Program Acquisition Cost By Weapon System: United States Department of Defense Fiscal Year 2021 Budget Request (2021), https://comptroller.defense.gov/Portals/45/Documents/defbudget/fy2021/fy2021_Weapons.pdf.

9. Irene Loewenson, "The Marine Corps Has Trained Its Final 8 Scout Snipers," Marine Corps Times, December 28, 2023, https://www.marinecorpstimes.com/news/your-marine-corps/2023/12/28/the-marine-corps-has-trained-its-final-8-scout-snipers/.

10. Pegler, *Sniper: A History of the US Marksman,* 149.

11. John Pike, "Napalm in War," GlobalSecurity.org, accessed August 4, 2022, https://www.globalsecurity.org/military/systems/munitions/napalm-war.htm#:~:text=During%20the%20Korean%20War%2C%20the,each%2C%20and%20held%20100%20gallons.

12. Daniel Byman et al., "Hamas's October 7 Attack: Visualizing the Data," CSIS, accessed April 22, 2024, https://www.csis.org/analysis/Hamass-october-7-attack-visualizing-data.

13. John Boyd and Grant T. Hammond, A Discourse on Winning and Losing § (1995), 384.

EPILOGUE

This book has been a labor of love, years in the making. I have chosen to write this in a formal manner, that is to say, I avoided using words like "I," "me," "we," etc. The result is that it reads somewhat like a textbook rather than a narrative. This was by design, as it is my wish to reach not only my fellow shooters, but also those within the elite academic circles and command staff. It is my hope that this text forces a paradigm shift, or at least provides momentum in that direction.

Not only do I want the public at large to have a better understanding of what we are, and who we are, but it is also an attempt to better educate future leaders within military and paramilitary operations and an attempt to inform them of what we do and what we can provide. I have tried my best to list those legends among us so as to garner interest and a larger audience. My own adventures, battles, and accomplishments, though they are important to me, pale in comparison. I know there is much more to write. I know that I did not cover everything. Please forgive me for what I missed. Much like our chosen profession, there are rules and guidelines within the world of an author.

Furthermore, like other historical offerings, this book sought to offer context in the framework of my own critical thinking. That is to say, although the facts are accurate, they are relayed here as I, a sniper, interpret them. I attempted to avoid what might be considered divisive rhetoric and, instead, tried to rely entirely on objective and universal truth. This is a difficult task with regards to interpreting human nature. Considering my own background in both the United States Army and as an American police officer, I like to believe that I could be considered a reasonable warrior. Perhaps that moniker needs some elaboration? Within American law enforcement, there exists the rules of criminal procedure. The rules of criminal procedure are concrete guidelines, scrutinized by the

U.S. Supreme Court, that must be followed in order to justify an officer's use of that power, which is granted to him/her by the community. For instance, the rules of criminal procedure define the "reasonable person," "reasonable suspicion," and "probable cause" standards that must be met in order to justify stopping, detaining, and/or arresting citizens. Furthermore, they define the "reasonable officer" standard for circumstances that involve the use of force upon the citizenry.

Though the profession of soldier and the profession of police officer resemble each other, and indeed share many commonalities, they often drastically differ from each other as well. For instance, there is an almost zero percent chance that a police officer will be justified in delivering suppressive fire in his/her capacity as an officer. Thus, instead of saying that I apply the reasonable officer standard to myself, insofar as the use of violence is concerned, I instead choose to apply a reasonable warrior standard, as I have been successful in both capacities.

Another way in which this text differs from traditional academic writing is with regard to my source material. Within formal historical academic writing, sources are divided into categories. The most coveted of these are primary sources, or sources that bore witness or were directly involved in the incidents being reported. Secondary sources are next, or those sources that have some expertise in what is being discussed but were not witness to said events. In this respect, many academics are great secondary sources. So, for example, in the early chapters of this book, citing historical documents like the Journals of the Second Continental Congress are primary sources while Martin Pegler would be an example of a secondary source. Although my methodology for citing sources does not differ from traditional academia, with respect to the utilization of primary and secondary sources, traditional academics often prefer to utilize only those sources found on paper. That is to say, if one wants to check their information, they often have to hunt down the book, journal, or historical document of which the reference was made. Much of my source material, on the other hand, is easily accessible via the internet. Certainly, much of my research utilized source material that can be found in the Library of Congress or within autobiographies of eras past, but a large amount can be found by navigating to a URL address. Nevertheless, much in the same way that I approach embracing the use of technology to better myself as a sniper, I chose to do the same in my approach to this book.

Finally, although my parting thought could be considered a shallow platitude, I offer the following: Embrace it. Be proud of it. Take all of the offered pejoratives and do what the American warrior class does best: Embrace the suck, laugh in its face and make it your own. After all, there are worse things to be called—like "weak."

APPENDIX A

ANATOMY OF A MODERN PRECISION RIFLE

The precision rifle is perhaps the most coveted individually operated weapon system in existence; yet most people are unfamiliar with its anatomy. Indeed, many believe it to be nothing more than a rifle with a scope attached, and it is often incorrectly referred to as a "sniper rifle." A more accurate nomenclature would be a sniper's rifle. It, after all, is only a precision rifle; the individual is the sniper. Without the user, the weapon system itself does not accomplish the mission. For instance, in the possession of an individual without the necessary education, training, and skill required, it is merely a heavy rifle, and an exceptionally slow one at that.

However, in the hands of one who is trained in its use, the precision rifle becomes an extremely effective tool. Nevertheless, the sniper's rifle varies based upon the mission. In the event that extreme precision is necessary, the sniper may carry a bolt-action heavy barreled rifle. In other situations where circumstances dictate the need for speed and maneuver, a gas operated rifle, such as an AR-10, may be the order of the day. In any event, this is meant to explain the elements required for what most imagine when using the term "sniper rifle."

This particular addition may prove provocative among snipers and precision shooters as there are often arguments centered around what is considered the most important elements of a precision rifle platform. Some may argue that the trigger mechanism is the most important while others will argue that it is the barrel of the weapon. In any event, the foundation of the discipline is consistency. During the marksmanship portion of training, potential snipers are trained to master fundamentals that lead to consistency from shot to shot. Thus, it follows logic that the weapon system itself be predicated on consistent shot placement. In order to achieve that, there are parts of the platform that differ from what might be considered an "ordinary" rifle. This chapter will seek to identify these differences and explain the various elements of a precision rifle.

Perhaps the most important difference between what might be considered an ordinary rifle and a precision rifle is the barrel. A typical hunting rifle must find a comfortable middle ground between precision, accuracy, and portability. That is to say that hunting rifles are often light and easy to shoulder should an opportunity to engage game present itself. Modern offerings feature composite material stocks that are significantly lighter than their wooden furniture counterparts. In the case of a precision rifle, however, a premium is placed on precision and accuracy while portability is of less concern. That is not to say the rifle is not man portable; rather that the function of the rifle dictates the necessity for increased weight. This is largely due to the heavy barrel, often referred to as a "bull barrel."

A bull barrel has a significantly thicker diameter than a standard rifle barrel and is responsible for most of the added weight. One of the reasons a heavier barrel is required is to mitigate the barrel vibration that occurs when a shot is fired. This vibration is known by different names, but is most commonly referred to as "harmonics," "whip," or "ringing." This can be likened to a soundwave, whereby a ripple effect occurs throughout the barrel when a shot is fired. When likened to an image of a soundwave, a heavier barrel essentially reduces the frequency (number of waves) and amplitude (height of waves), while increasing the wavelength (distance of waves from peak to peak.) It also increases the consistency of the waves generated, meaning that the barrel vibrates nearly the same way with each shot, although that is also predicated on the consistency of the ammunition being used. Another variable that mitigates is heat transfer. As the steel in a barrel increases in temperature, the molecular structure is altered, even if only an imperceptible amount.

Each time a shot is fired, the expanding gas of the round is generating an enormous amount of heat and pressure. In addition, as the projectile is forced down the barrel, the friction of the projectile against the lands and grooves is adding to this increase in temperature. Heavier barrels do not heat up nearly as quickly as standard barrels, thus consistency from shot to shot is easier to maintain. Finally, the added weight of the heavy barrel decreases the amount of felt recoil, reducing the amount of "kick" generated by the rifle. Within the world of marksmen, mitigating felt recoil increases the marksman's ability by reducing what is known as "anticipation of recoil," whereby the marksmen will flinch, or slightly dip the muzzle just prior to breaking a shot as an attempt to compensate for the "muzzle flip" generated by the recoil.

Anticipation of recoil is one of the most difficult habits to break and shooters can train for years before completely eliminating it. Within the selection of types of heavy barrels, there are many contours, or "profiles," to choose from. The nostalgic M-24 Sniper Weapon System of the United States Army features a straight profile barrel, meaning that the diameter of the barrel at the chamber is the same as the diameter of the barrel at the muzzle. As one can imagine, this makes the rifle significantly heavier.

Different barrel manufacturers will often give proprietary names to barrel profiles. For instance, the Remington 700 offers a barrel profile known as the "Sendero," where the barrel tapers down in size from chamber to muzzle. Other manufacturers may have the same profile under a different name. Tapered barrel profiles are a manner by which the manufacturer can maintain heavy barrel integrity while mitigating increases in weight. Other manufacturers produce "fluted" barrels, where channels are carved into the exterior of the barrel, maintaining heavy barrel integrity while mitigating weight and increasing the dissipation of heat. Essentially it boils down to the shooter's preference and ammunition. Some shooters swear by one barrel profile whereas others swear by another. This text does not seek to identify which is better, but merely to inform the reader as to what is meant by the term *barrel profile*.

Insofar as bolt-action rifles are concerned, the next most obvious characteristic is what shooters refer to as the "action." The action is the upper receiver of the rifle and the bolt assembly. In a bolt-action, each round is manually fed into the chamber via operation of the bolt handle. This differs from automatic and semi-automatic weapons, whether gas or recoil operated, as the spent cartridge is not automatically ejected and a fresh cartridge inserted into the chamber via an automatically reciprocating bolt. That is not to say that only a bolt-action rifle can be utilized as a precision weapon system. Many AR platform gas operated rifles are upgraded with bull barrels and free floated handguard assemblies making them effective precision weapon platforms. However, arguments abound as to whether or not a gas operated rifle is capable of being as precise as a bolt-action.

The argument hinges on the gas impingement system and the ability to control how much/how little gas is allowed to escape into the gas tube to cycle the bolt carrier group, whereas a bolt-action—being manually charged after each shot—eliminates that variable.

With advances in technology, it is plausible that a gas operated rifle could be manufactured to such strict tolerances that it is capable of producing the same precision as a bolt-action. However, a definitive answer to this debate may never come to fruition as the variables, to include the burn rates of powder, are impossible to perfectly—intentionally—replicate from shot to shot. It is safe to say that, as of the penning of this text, c. 2024, Minute of Angle accuracy is easily accomplished by many gas operated precision rifles being manufactured. Perhaps the debate continuing into sub-minute of angle precision is superfluous and literally splitting hairs.

Essentially, the action feeds the ammunition into the chamber of the barrel and removes spent cartridges. Perhaps the most well-known action utilized today is the Remington 700. It is utilized by snipers in the United States Army, Air Force, Navy, and Marine Corps as well as, many police SWAT teams, the Secret Service, and the Federal Bureau of Investigation H.R.T. For the purpose of this text, focus will be given to the M-24 of the United States Army and the M-40 of the United States Marine Corps as all branches of the military utilize one or the other. The major difference between the two is the length of the action. Though both branches have initially chambered their rifles to accept the 7.62 x 51mm NATO (.308 Caliber) cartridges, the Army utilizes what is known as a "long-action," whereas the Marine Corps uses a "short-action." The difference between the two is the variety of ammunition each can accept. The M-24's long-action means that, by changing the barrel, the rifle can accept longer cartridge ammunition, specifically the .300 Winchester Magnum.

The Marine Corps' various iterations of the M-40, being a short-action, means that the .308 is the largest cartridge it will hold. Notwithstanding, both, with a barrel change, can accept shorter cartridges, the 6.5mm Creedmoor for example. Insofar as appearance and/or functionality is concerned, the long-action is just as its name suggests, longer in length than a short-action. In 2017, at the Aberdeen Proving Ground, the United States Army began testing other cartridges for precision rifles. The tried and true .308 was matched against the 6.5mm Creedmoor and the .260 Remington cartridges. The 6.5mm Creedmoor met or exceeded all expectations during the course of testing.

In 2018, the Hornady cartridge company was awarded a Department of Defense contract to supply 6.5mm Creedmoor ammunition, which it initially developed in 2008, as an intermediate range sniper cartridge.[1] In 2018, Special Operations Command (SO-

---

1. J. Scott Rupp, "Hornady Awarded 6.5 Creedmoor Military Contract," RifleShooter (RifleShooter, May 1, 2020).

COM) began their own testing, pitting the .260 against the 6.5mm Creedmoor[2], thus it may turn out that the 6.5mm Creedmoor becomes the preferred intermediate range sniper cartridge throughout U.S. military sniper programs. Aside from ejecting a spent cartridge and inserting a new one, the bolt itself has a number of lugs that, once the bolt is rotated—cocking the weapon—hold the bolt firmly in the receiver.

The Remington 700, as well as many of its clones, utilizes two large lugs to accomplish this. This firm lock up in battery allows the use of larger loads, as the chamber and the lugs themselves can withstand higher pressures than can a lever-action for example. Although the number of lugs on the bolt may vary among manufacturers, the principle is the same. The more precisely they are machined, the closer they fit and the tighter they hold. Thus relegating a bolt-action receiver to merely expending spent cases and inserting fresh cartridges is an oversimplification.

Next is the stock or chassis of the system. Rifles of the past featured wooden stocks in which the action could be bolted. The M1 Garand, Mosin-Nagant, and Gewehr 98 are all classic examples of combat rifles that were converted to meet the needs of snipers. Each was utilized by line troops on the front and were retrofitted with accoutrements to facilitate optics after the fact. Though they were accurate enough to satisfy the role, as technology advanced and ballistics became better understood, the wooden furniture of the rifle was continuously reduced. Soldiers returning home from war would keep their rifles and "sporterize" them for use in hunting. Essentially, they removed as much of the wooden stock as they could to make the weapon lighter and easier to carry while hunting or target shooting. The result, albeit unintentional, would later become known as "free-floating" the barrel.

The term free-floated barrel means that the barrel does not come into contact with the stock past the chamber of the rifle. This allows the barrel vibrations, or harmonics, to react naturally and consistently from shot to shot. For this reason, snipers will avoid resting their barrel on objects to steady their shots and will instead rest the stock on an object, such as a backpack, or utilize a mounted bipod that is attached to the stock/chassis to stabilize the rifle.

Eventually, the wood of the stock was eliminated. This was done for a number of reasons, the first of which was that composite materials such as plastics and alloys were

---

2. Todd South, "Special Operations Command Is Looking at a New 6.5 Mm Round for Its Sniper Rifle," *Military Times* (Military Times, August 19, 2022).

manufactured to be stronger and lighter than wood. Secondly, wooden stocks are suscep-tible to the elements. A sniper's rifle, although a precision instrument, is constructed to be stout and capable of handling harsh environments and treatment while still producing precise results. Accordingly, the rifle is subjected to rain, snow, mud, dirt, sand, and anything else that mother nature can throw at it. Over time, a wooden stock that is subjected to this, especially wet conditions, will warp. Although a warp in the wood can be corrected for, if it warps to a point whereby the wood touches the barrel, it eliminates a free-floated barrel and decreases the precision and accuracy of the rifle.

Plastics, alloys, and carbon fibers, on the other hand, are less susceptible to the elements than wood. The M-24 SWS of the Army features a stock produced by H-S Precision. The H-S Precision stock is fabricated from aramid fiber reinforced fiberglass and features an aluminum block at its core. This makes the stock resilient to almost all environmental conditions. The aluminum block at its core serves two purposes. The recoil lug, a piece of steel that is attached to the barrel where it meets the action, fits snuggly against this aluminum alloy block. The lug transfers energy from the recoil into the stock, allowing that energy to spread over a larger surface area as the rifle recoils into the shooter.

Secondly, the bolts that connect the stock to the receiver are fed through this aluminum block. In wood stocks, these holes would get damaged over time due to recoil. Shooters of the past would mitigate damage to the wood blocks by "glass and pillar bedding" the block of the wooden stock. During this process, liquid epoxy or fiberglass would be poured into the inlet of the stock. The action of the rifle would then be greased, seated, and bolted into the stock until the epoxy dried. Upon drying, the stock was then bedded precisely to the action, and the block of the stock was protected from damage caused by recoil.

Most modern offerings for precision rifles only utilize composite materials or alloys in the construction and manufacture of their stocks and chassis, thus eliminating these problems of the past. The obvious question follows: "What is the difference between a chassis and a stock?" This is a fairly simple question with a complicated answer. Technically, the difference—aside from outward appearance—is the manner by which they are constructed. A stock is often constructed out of wood or composite materials; afterward, the block in the inlet is reinforced. Sometimes, a stock will be constructed around a stronger material block. A chassis, on the other hand, is machined out of one piece of aluminum alloy. Thus, there is no need to secure a separate block within it. Notwithstanding, chassis can have other accoutrements attached to it. The Magpul Pro

700, for instance, is a chassis system with plastic furniture attached for aesthetic and modular purposes.

Another popular series of chassis systems is manufactured by Masterpiece Arms, often shortened to MPA. That is not to say that a chassis is superior to a stock. Some stock manufacturers, such as the Manners Composite Stocks company manufacture stocks that are equally substantial to a chassis. In the end, it once again comes down to an individual shooter's preference. Stocks have a more nostalgic appearance whereas chassis systems look more 'space age.' Modern offerings of both feature adjustable combs, the portion of the butt stock on which the shooter rests his/her face, and an adjustable length of pull. The length of pull is the length of the butt stock, which can be increased or decreased to better suit the shooter. These adjustable features allow the shooter to customize the rifle to his/her specific body type, making each rifle customized specifically to the shooter.

Finally, the most prominent feature of the precision rifle system is the scope. Indeed, an entire chapter—or book—could be dedicated to the sniper's scope and there are far too many variations of telescopic rifle sites to cover them all adequately within this text. Thus, attention will be focused on modern sniper scopes that have been utilized in recent conflicts. The premise of a scope is simple, magnified vision with a crosshair, or reticle, to provide a point of aim.

It is theorized that a rudimentary scope may have been utilized by the shooter who killed General Simon Fraser at the Battle of Bemis Heights near Saratoga, New York in 1777. However, as technology became more advanced, so too did telescopes. The modern scopes of today only loosely resemble their predecessors. The basic breakdown of the modern scope, from front to back, consists of a large objective lens, a scope tube, graduated turrets, a magnification adjustment, a focus adjustment, and an ocular lens. The diameter of the scope tube and various lenses can differ by scope type and from manufacturer to manufacturer.

Perhaps the most recognizable feature of the modern scope are the large turrets that protrude from the center of the scope body. These turrets allow for the sniper to make measurement adjustments depending on the range of their target. The top turret is the elevation adjustment turret while the side turrets are typically parallax correction on the left and windage adjustment on the right. While the parallax adjustment corresponds to focal planes, the elevation and windage turrets correspond directly to the reticle of the scope. Parallax is perhaps the most difficult to explain. Merriam-Webster's dictionary defines it as "the apparent displacement or the difference in apparent direction of an object as seen from two different points not on a straight line with the object."[3]

A more familiar analogy of the phenomenon of parallax is to imagine looking out of the side window of a vehicle while travelling. The trees nearby seem to pass far more quickly than do the mountains in the background. Simply stated, adjusting the parallax ensures that the reticle matches the correct focal plane of the target. Windage and elevation are somewhat easier concepts to explain. Both turrets utilize systems of angular measurement to correct for external ballistics. In the past, the reticle utilized milliradians while the turrets utilized minutes of angle. Based upon what the sniper observed and measured via his/her reticle, he/she would then execute a mathematical conversion to make the corresponding adjustment. A minute of angle is defined as $1/60^{th}$ of one degree, or approximately 0.016 degrees. A radian is 57.3 degrees, thus one milliradian is $1/1000^{th}$ of a radian, or 0.0573 degrees. Notwithstanding, snipers do not pay much attention to what each represents in a circle, but more particularly, what each represents in the scope. That is to say, the sniper simplifies it. To a precision shooter, one minute of angle equals approximately one inch (1.047 inches) at 100 yards while one milliradian equals 3.6 inches at 100 yards. Because the measurements are angular, the measurement increases with the distance by a factor of ten. Thus, one minute of angle equals approximately two inches at 200 yards, three inches at 300 yards and so on. The milliradian system is better expressed utilizing the metric system, whereby one milliradian at 100 meters is equal to 10 centimeters and at 200 meters it is equal to 20 centimeters etc.

Nevertheless, it still translates well enough into the imperial system of measurement. Simply put, there are approximately three minutes of angle in one milliradian. Notwithstanding, this is not the vernacular used by snipers and precision marksmen on the line; rather the terms are truncated to "minute" and "mil." Snipers of the past utilized what is known as the "mil-dot" reticle, in which there was exactly one milliradian between the center of each dot in the reticle and there were four dots extending from the center on each crosshair.

This reticle could be used to measure distant targets to apply a range estimation formula or calculate wind holds or holds for moving targets. The corresponding turrets graduated in quarter minute of angle increments, thus a one mil discrepancy in the reticle required at least a three-minute correction on the corresponding turret, which

---

3. "Parallax Definition & Meaning," Merriam-Webster.

manifests as twelve "clicks." It quickly becomes apparent how confusing this calculation and conversion becomes, which is why snipers will train for years to master it.

At the beginning of the twenty-first century, a change began taking place within precision optics. Manufacturers, in the interest of innovation, began simplifying the scopes. Instead of mil-dot reticles combined with M.O.A. turrets, the manufacturers began matching the units of measurement of the turrets to the reticle. Thus, a mil reticle had matching mil turrets graduated in one tenth increments, known as a "mil-mil" or "MRAD" scope, eliminating the need to perform a mathematical conversion prior to a shot correction. Other scopes makers changed the reticle to match the turrets, creating an M.O.A. reticle as opposed to a mil reticle, accomplishing the same end. The reticles themselves also saw innovation. The dots utilized in the past were usurped by much finer mil or minute hashes, making measurements far more precise.

In addition, ballisticians began calibrating reticles for windspeed and moving target corrections, allowing the shooter to utilize a small dot in the reticle for predetermined windspeeds at distance, eliminating the need to dial the windage turret. This created what many shooters refer to as the "Christmas tree" reticle, where the lower half of the reticle resembles a Christmas tree shape. A fine example of this is the Tremor 3 reticle, created by the Horus company, which is now featured in many high-end scopes, to include the Leupold Mk 5-HD, the Kahles K525i, the Night Force ATACR, and the Steiner M7XI. Although some prefer less "junk" in the reticle, these innovative reticles are gathering steam and will only become more advanced as time goes by.

At the time of the writing this text, the technology utilized to increase effectiveness and lethality is increasing exponentially. Though warfare is a motivator insofar as this innovation is concerned, it is the civilian market that is the driving force behind this concept. That is not a new dynamic. After all, the civilian market Lyman Alaskan was the scope utilized on many Garand M-1 sniper rifles. Nevertheless, modern telescopic optics would have seemed to come directly out of a Jules Verne novel to snipers in World War II. One modern scope, in particular, the Vortex NGSW-FC (Next Generation Squad Weapon—Fire Control,) combines multiple components into a single (relatively compact) system. Instead of just a magnified telescopic optic, the NGSW-FC features "a variable magnification optic, backup etched reticle, laser rangefinder, ballistic calculator, atmospheric sensor suite, compass, Intra-Soldier Wireless, visible and infrared aiming lasers, and a digital display overlay."[4]

Essentially, the scope does all of the work for the sniper, to include designating a point-of-aim within the field of view. That is not to say that the sniper selection process or training regimen should be eliminated or relaxed. While technology should be embraced and implemented, it is also apt to fail. Electronics overheat, circuit boards short circuit, and batteries die. The addition of "gizmos" and "thingamajigs" should only ever append to mastered fundamentals, hence the featured "backup etched reticle." It is safe to assume that after an investment of that magnitude, the U.S. Army, at least, does not have any intentions of eliminating its sniper program and is instead dedicated to bettering it.

4. Todd South, "Army Times," *Army Finally Picks an Optic for Next Generation Squad Weapon*, January 7, 2022, https://www.armytimes.com/news/your-army/2022/01/07/army-finally-picks-an-optic-for-next-gene ration-squad-weapon/.

APPENDIX B

THE OODA LOOP

There are two ways to defeat a sniper . . . the first requires employing a large amount of explosive ordnance directly, by air, or indirectly, via artillery. This tends to generate unwanted collateral damage. The second, and far less destructive, way to engage an enemy sniper is with another sniper, or "counter-sniper." Counter-sniper operations are difficult and dangerous. There is a desire to attribute a successful counter-sniper operation to luck, and perhaps that may be the case, but skill and experience play a significant role. Thus, it stands to reason that if an adversary is utilizing snipers in their warfare doctrine, snipers must also be implemented by friendly forces. That does not necessarily mean that a friendly sniper must immediately "kit-up" and go hunt for the enemy sniper. It could mean that the friendly sniper assists in a tactical operations center, by identifying possible hide locations on a sector map for subsequent indirect fire. Regardless, only a trained sniper is capable of thinking like a trained sniper.

There are a number of training doctrines that support this, but perhaps the concept that best illustrates this is known as the OODA Loop. The word "OODA" is an acronym for Observe, Orient, Decide, Act and is designed to explain the feedback loop that is processed by the brain when encountering a scenario.[1] The concept was developed by

---

1. John Boyd and Grant T. Hammond, A Discourse on Winning and Losing § (1995), 383.

an American Air Force fighter pilot named Colonel John Boyd who served in World War II, the Korean War, and the Vietnam War.

During that time, Boyd generated what is known as a "disruption" strategy in order to defeat enemy fighters in air-to-air combat. This concept went on to become a strategy utilized not only in warfare, but in business and litigation as well. Simply explained, one must observe a scenario, orient themselves to the scenario, decide what to do, then act on that decision. Colonel Boyd's theory is that by disrupting this feedback loop, one could cause confusion and instability in an adversary, thereby providing an opportunity to strike. This theory, however, is predicated on experience and training. Therefore, one would not send an artillery officer into an aerial dogfight. That is not to say that one is more important than the other.

Both are necessary for the prosecution of successful warfare campaigns, but only because both have expertise in their respective fields. Furthermore, one's own OODA loop must also be trained to reduce the amount of time it takes to process each step, thereby decreasing one's response time and increasing their effectiveness to an adversary's own disruption techniques. Thus, a concept that is somewhat easy to describe becomes incredibly difficult to successfully implement. In short, until all adversaries decide, collectively, to discontinue the use of snipers, snipers will remain necessary. Considering that it is universally accepted that the use of nuclear weapons would result in the catastrophic downfall of humankind, and, that the use of nuclear weapons is still a clear and present danger, it is unlikely that the nations of the world, let alone terrorist organizations and their ilk, will collectively decide to eliminate the use of snipers.

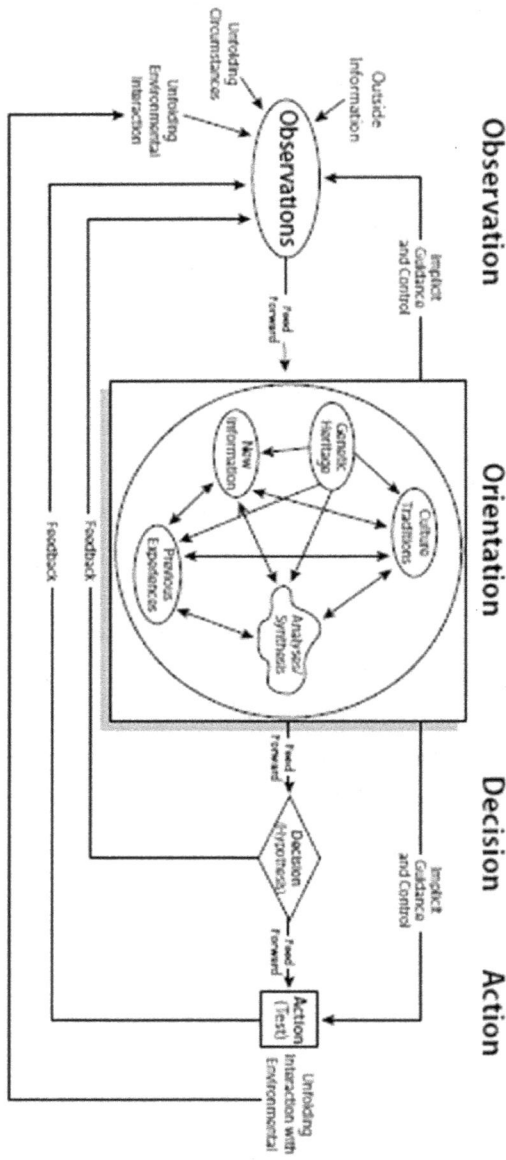

The OODA Loop. Boyd's final sketch of the OODA Loop, as presented in his summation of "A Discourse on Winning and Losing," which he referred to as "the big squeeze," 28 June 1995. Adapted from Hammond, *Mind of War*, 190.

John Boyd and Grant T. Hammond, A Discourse on Winning and Losing, 384

APPENDIX C

DEFINITIONS

Action: The functional mechanism of a firearm that manipulates (loads, locks, fires, extracts, and ejects) ammunition.

Ballistics: The science concerning the mechanics of firearms and projectiles.

- **External Ballistics:** The characteristics of a projectile in flight.

- **Internal Ballistics:** The characteristics of the chemical reaction occurring as a round of ammunition is detonated within the chamber of a firearm. Additionally, it is the characteristics of the projectile within the barrel of a firearm.

- **Terminal Ballistics:** The characteristics of a projectile once it strikes an intended target.

Ballistic Coefficient: The measure of a projectile's ability to mitigate air resistance.

Bipod: A two-legged stand that attaches to the stock/chassis of a rifle.

DOPE: Data On Previous Engagements. Data that is maintained by a sniper that charts the characteristics of shots taken in as many situations as possible, training or otherwise.

Favor: Placing the center of the point-of-aim half way between the center and the extreme edge (left, right, top, bottom) of the intended target. Colloquially known as

"Kentucky Windage." Utilizing the precision measurements of the reticle, as opposed to the turrets, to make elevation or windage adjustments.

Example: "Favor right and send it." This tells the sniper to move the center of the crosshair half way between the center and the right edge of the target and fire.

Hold: Placing the center of the point-of-aim on the extreme edge (left, right, top, bottom) of the intended target. Colloquially known as "Kentucky Windage." Utilizing the precision measurements of the reticle, as opposed to the turrets, to make elevation or windage adjustments.

Example: "Hold left and send it." This tells the sniper to move the center of the crosshair to the left edge of the target and fire.

HVT: High Value Target

Loophole: A small hole, manmade or otherwise, that provides a vantage point from which a sniper can fire with relative impunity.

MOA: Minute of Angle—A unit of angular measurement that is equal to $1/60^{th}$ of 1 degree. In imperial measurements, this is equal to 1.047 inches at 100 yards and usually approximated to 1 inch @ 100 yards. The measurement is proportional as its imperial measurement increases relative to distance whereby 1 minute of angle is equal to approximately 1 inch at 100 yards and approximately 10 inches at 1,000 yards.

Mil-Dot: A dot on a precision site reticle. In Leupold model scopes, it is a circle that is 1 moa in width x 1 moa in height. The center of each dot is situated 1 milliradian apart.

Milliradian/Mil-Rad/MRad: A unit of angular measurement representing $1/1000^{th}$ of a radian. This is equal to 10 centimeters or 3.6 inches at 100 yards. Usually measured in tenths of radians for precision adjustments.

Musket: A man-portable, shoulder fired, firearm with a long, smooth-bored, barrel.

Parallax: The effect whereby the position or direction of an object appears to differ when viewed from different positions, e.g. a rifle scope.

Precision Rifle: A rifle capable of 1 minute of angle precision and accuracy at 100 yards.

Radian: A unit of angular measurement equal to 57.3 degrees.

Reticle: A series of fine lines or fibers in the eyepiece of an optical device, such as a rifle scope, used as a measuring scale or an aid in point-of-aim.

Rifle: A man-portable, shoulder fired, firearm with a long, rifled, barrel featuring lands and grooves that impart a gyroscopic spin on a fired projectile.

Sniper: An individual, usually a member of a military or police force, who is adept at advance camouflaging practices and reconnaissance, and can deliver precise and accurate rifle fire that may exceed the range of standard infantry capabilities.

Spotter: A member of a sniper team, usually the more senior sniper, who utilizes a spotting scope or other means of observation to provide real time information and environmental data to a sniper.

Spotting Scope: A high power optic utilized for surveillance and watching environmental conditions, typically utilized by a spotter.

Stock: A device to which a barrel and an action are fitted, that enables a shooter to wield, support, and fire a long-barreled firearm. Usually made out of wood, but can also be constructed of more modern materials such as plastic or carbon fiber.

SVD: *Snayperskaya Vintovka Dragunova* (Russian for Dragunov Sniper Rifle.) Soviet era gas impingement system, semi-automatic sniper rifle, designed between 1958–1963. It features the AK, *Avtomat Kalishnikova* (Russian for Kalishnikov's Automatic) receiver.

Trajectory: The parabolic arc, also known as a flight path, of a projectile.

Turret: A knob on a telescopic optic utilized for adjustments.

- **Elevation Turret:** A knob, usually on the top of a rifle scope, that adjusts the rifle's point of aim, allowing for the Y-Axis adjustments (MOA or Mil) required for trajectory.

- **Parallax Turret:** A knob, usually on the left side of a rifle scope, that adjusts the parallax focus.

- **Windage Turret:** A knob, usually on the right side of a rifle scope, that adjusts the rifle's point of aim, allowing for the X-Axis adjustments (MOA or Mil) required for negotiating wind or moving targets.

BIBLIOGRAPHY

"9/11 Faqs." 9/11 FAQs | National September 11 Memorial & Museum. Accessed September 6, 2022. https://www.911memorial.org/911-faqs.

Adams, John. *The John Adams Papers*. Edited by Robert J Taylor. Cambridge, MA: Harvard University Press, 1979.

"Advanced Infantry Training Battalion." Official Website of the United States Marine Corps. United States Marine Corps. Accessed November 4, 2020. https://www.trngcmd.marines.mil/Units/South-Atlantic/SOI-E/Units/Advanced-Infantry-Training-Battalion/Scout-Sniper-Course/.

Afong, Milo S. *Hunters: U.S. Snipers in the War on Terror*. New York, NY: Berkley Caliber, 2013.

Armstrong, Nevill A. D. *FIELDCRAFT, SNIPING AND INTELLIGENCE*. Reprinted. E. Sussex: Naval & Military Press, 2019.

Asymmetric Warfare Group, Russian New Generation Warfare Handbook § (2016).

Beevor, Antony. *Stalingrad*. London: Penguin Books, 2017.

"Beltway Snipers." FBI. Federal Bureau of Investigation, May 18, 2016. https://www.fbi.gov/history/famous-cases/beltway-snipers.

"Berdan, Colonel of the Sharpshooters." *White Cloud Kansas Chief*, December 26, 1861. https://www.loc.gov/resource/sn82015486/1861-12-26/ed-1/?sp=1&r=0.642,0.801,0.29,0.173,0.

Berdan, Hiram. "SHARPSHOOTERS OF THE NORTH." *The New York Herald*. May 31, 1861, Morning edition. https://www.loc.gov/resource/sn83030313/1861-05-31/ed-1/?q=Hiram+Berdan+Sharpshooters&sp=5&st=text&r=0.498,0.995,0.181,0.23,0.

"Berdan's Regiment of Sharpshooters - Examination of Applicants at Wee-hawken." *Chicago Daily Tribune*, July 27, 1861, Volume XIV edition, sec. Number 324. https://www.loc.gov/resource/sn84031490/1861-07-27/ed-1/?sp=1&r=0.04 8,0.041,0.746,0.446,0.

"Berdan's Regiment of Sharpshooters - Interesting Examination of Applicants at Wee-hawken." *New York Herald*. July 24, 1861. https://www.loc.gov/resource/sn83030313 /1861-07-24/ed-1/?sp=8&st=text&r=0.051,0.009,0.417,0.499,0.

Blehm, Eric. *Fearless: The Undaunted Courage and Ultimate Sacrifice of Navy SEAL Team Six Operator Adam Brown*. Colorado Springs, CO: WaterBrook Press, 2017.

Bowden, Mark. *Blackhawk Down: An American War Story*. Philadelphia, PA: Philadelphia Inquirer, 1997.

Boyd, John, and Grant T. Hammond, A Discourse on Winning and Losing § (1995).

Bush, George W. Report, Address to a Joint Session of Congress and the American People § (2001).

Byman, Daniel, Riley McCabe, Alexander Palmer, Catrina Doxsee, Mackenzie Holtz, and Delaney Duff. "Hamas's October 7 Attack: Visualizing the Data." CSIS. Accessed April 22, 2024. https://www.csis.org/analysis/Hamass-october-7-attack-visualizing-da ta.

"Col. Berdan's Sharpshooters." *New York Times*. August 7, 1861.

Culbertson, John J. *13 Cent Killers: The 5th Marine Snipers in Vietnam*. New York: Ballantine Books, 2003.

Delegation of Ukraine, Statement by the Delegation of Ukraine at the 947th FSC Plenary Meeting on Russia's ongoing aggression against Ukraine and illegal occupation of Crimea § (2020).

Durant, Michael J, and Steven Hartov. *In the Company of Heroes:* New York City, NY: New American Library, 2003.

Fehrenbach, T. R. *This Kind of War: The Classic Military History of the Korean War*. New York, NY: Open Road Integrated Media, 2014.

Ford, Worthington C., ed. *Journals of the Continental Congress: 1774-1789*. 2. Vol. 2. Washington: Government Print Office, 1904. https://babel.hathitrust.org/cgi/pt?id=u c1.31822019445410&view=1up&seq=93.

Ford, Worthington Chauncey, Gaillard Hunt, John Clement Fitzpatrick, Roscoe R. Hill, Kenneth E. Harris, and Steven D. Tilley, *Journals of the Continental Congress*, 1774-1789 § (1904).

https://memory.loc.gov/cgi-bin/ampage?collId=lljc&fileName=002/lljc002.db&recNu m=88&itemLink=r?ammem/hlaw:@field(DOCID+@lit(jc00235))%230020090&link Text=1.

Gates, Horatio. "Founders Online: To George Washington from John Hancock, 30 September 1777." National Archives and Records Administration. National Archives and Records Administration. Accessed November 3, 2020. https://founders.archives. gov/documents/Washington/03-11-02-0378.

——— "To George Washington from Brigadier General Horatio Gates, 22 June 1775." National Archives and Records Administration. Accessed November 3, 2020. https://founders.archives.gov/documents/Washington/03-01-02-0011.

Gatopoulos, Derek, and Adam Pemble. "Volunteer Sniper Embodies Ukraine's Versatile Military." AP NEWS. Associated Press, August 30, 2022.

Hedges, Chris. "War in the Gulf: The Marines; War Is Vivid in the Gun Sights of a Sniper." *The New York Times*. February 3, 1991.

Henderson, Charles. *Marine Sniper: 93 Confirmed Kills*. New York: Berkley Caliber Books, 2005.

Hesketh-Prichard, H. *Sniping in France*. Barnsley: Pen & Sword Military, 2014.

Higginbotham, Don. *Daniel Morgan Revolutionary Rifleman*. Chapel Hill: The University of North Carolina Press, 2014.

"History: Army Sniper Association." Army Sniper Association | Army Sniper Association, June 2, 2017. https://www.armysniperassociation.org/about/history/.

Hunt, Ira Augustus. *The 9th Infantry Division in Vietnam: Unparalleled and Unequaled*. Lexington, KY: University Press of Kentucky, 2010.

Hunt, Lynn. "Against Presentism." *Perspectives on History*. The American Historical Association. May 1, 2002. .

I., J. L. "Palmetto Sharp Shooters." *Yorkville Enquirer*, September 10, 1862. https://www.loc.gov/resource/sn84026925/1862-09-10/ed-1/?sp=2&r=-0.07,0. 712,0.84,0.502,0.

Johncox, B. James. Author's Interview with James Gilliland. Personal, October 20, 2020.

Kaonga, Gerrard. "Top Ukraine Sniper Compares Taking out Russians to Going on Safari." Newsweek. Newsweek, May 20, 2022. https://www.newsweek.com/ukraine-sn iper-olena-bilozerska-safari-russia-war-1708178.

Keller, Jared. "The US Military Is Losing the Sniper War Against Russia." Military.com, July 9, 2020. https://www.military.com/daily-news/2020/07/09/us-military-losing-sniper-war-against-russia.html.

Kyle, Chris, Scott McEwen, and Jim DeFelice. *American Sniper: The Autobiography of the Most Lethal Sniper in U.S. Military History*. New York: William Morrow Paperbacks, 2017.

Lacz, Kevin, Ethan E. Rocke, and Lindsey Lacz. *The Last Punisher: A SEAL Team Three Sniper's True Account of the Battle of Ramadi*. NY, NY: Threshold Editions, 2017.

Lanning, Michael Lee. *Inside the Crosshairs: Snipers in Vietnam*. New York: Ballentine, 1998.

Luscombe, Belinda. Chris Kyle: American Sniper | 10 Questions | TIME. Other. *TIME*, 2011. https://www.youtube.com/watch?v=aJ12PN81xnI.

Marcot, Roy M. *U.S. Sharpshooters: Berdan's Civil War Elite*. Mechanicsburg, MA: Stackpole Books, 2007.

Martin, Glen E. "They Call Their Shots." *Marine Corps Gazette* 37, no. 4, April 1953.

McMahon, Martin. *Sedgwick Memorial Association*. Philadelphia, PA: Dunlap & Clarke Printers, 1887. https://www.battlefields.org/sites/default/files/atoms/files/Sedgwick%20Monument%20Dedication%20at%20Spotsylvania.pdf.

"Military Advisors in Vietnam: 1963." Military Advisors in Vietnam: 1963 | JFK Library. Accessed August 19, 2022. https://www.jfklibrary.org/learn/education/teachers/curricular-resources/high-school-curricular-resources/military-advisors-in-vietnam-1963.

Mitchell, James L. *Colt: A Collection of Letters and Photographs about the Man, the Arms, the Company*. Stackpole, 1959.

Office of the Under Secretary of Defense (Comptroller)/Chief Financial Officer, Program Acquisition Cost By Weapon System: United States Department of Defense Fiscal Year 2021 Budget Request (2021). https://comptroller.defense.gov/Portals/45/Documents/defbudget/fy2021/fy2021_Weapons.pdf.

Pavlychenko Liⱆudmyla Mykhaïlivna, Martin Pegler, David Foreman, and Alla Begunova. *Lady Death: The Memoirs of Stalin's Sniper*. Newport, NSW: Big Sky publishing, 2018.

Pegler, Martin, and J. D. Penney. Personal Correspondence with Corporal J. D. Penney. Other. *Sniper: A History of the US Marksman*, October 2011.

Pegler, Martin. *Out of Nowhere: A History of the Military Sniper, from the Sharpshooter to Afghanistan*. Oxford: Osprey, 2011.

———. *Sniper: A History of the US Marksman*. Oxford: Osprey Publishing, 2011.

———. *Sharpshooting Rifles of the American Civil War*. Oxford: Osprey Publishing, 2017.

———. *Sniping in the Great War*. Barnsley: Pen & Sword Military, 2017.

———. *Sniping Rifles on the Eastern Front 1939-45*. Oxford: Osprey Publishing, 2019.

———. "The Allies Strike Back: The Genesis of Sniping, Part 5." An Official Journal Of The NRA, May 25, 2017. https://www.americanrifleman.org/content/the-allies-st rike-back-the-genesis-of-sniping-part-5/.

———. *The Military Sniper since 1914*. Oxford: Osprey Publishing, 2002.

Petain, Phillipe. Letter to Armies of the North and the Northeast. "Training of American Units with the French." *United States Army in the World War, 1917-1919*. University of California, Berkeley, May 1, 1989. .

Petraeus, David, and Jake Tapper. Picking them off: Petraeus explains how Ukrainians are taking out Russian generals. Other. CNN, March 20, 2022.

Pike, John. "Napalm in War." GlobalSecurity.org. Accessed August 4, 2022. https://www.globalsecurity.org/military/systems/munitions/napalm-war.htm#:~:text= During%20the%20Korean%20War%2C%20the,each%2C%20and%20held%20100%20g allons.

Plaster, John L. "Sniping in Ukraine." An Official Journal Of The NRA. Accessed October 21, 2022. https://www.americanrifleman.org/content/sniping-in-ukraine/.

———. *Ultimate Sniper: The Video*. *Youtube.com*. Paladin Press, 2020. https://www .youtube.com/watch?v=BIUdsU6y1i4.

Pyle, Ernie. *Brave Men*. Reprinted. Lincoln, NE: University of Nebraska Press, 2001.

Roberts, Craig, and Charles W. Sasser. *Crosshairs on the Kill Zone: American Combat Snipers, Vietnam through Operation Iraqi Freedom*. New York: Pocket Star Book, 2004.

Rupp, J. Scott. "Hornady Awarded 6.5 Creedmoor Military Contract." RifleShooter. RifleShooter, May 1, 2020. https://www.rifleshootermag.com/editorial/hornady-awar ded-65-creedmoor-military-contract/375847.

Sasser, Charles W., and Craig Roberts. *One Shot, One Kill*. New York: Pocket Books, 1990.

Senich, Peter R. *U.S. Marine Corps Scout-Sniper: World War II and Korea*. Boulder, CO: Paladin Press, 1993.

"Sniper Definition & Meaning." Merriam-Webster. Merriam-Webster. Accessed September 5, 2022. https://www.merriam-webster.com/dictionary/sniper#other-words.

"Sniper." sniper noun - Definition, pictures, pronunciation and usage notes | Oxford Advanced Learner's Dictionary at OxfordLearnersDictionaries.com. Oxford Learner's Dictionary. Accessed September 5, 2022. https://www.oxfordlearnersdictionaries.com/us/definition/english/sniper?q=Sniper.

Snow, Shawn. "The Sniper Shortfall: Why the Corps Could Lose Its next Urban Fight." Marine Corps Times. Marine Corps Times, November 13, 2018. https://www.marinecorpstimes.com/news/your-marine-corps/2018/11/13/the-sniper-shortfall-why-the-corps-could-lose-its-next-urban-fight/.

South, Todd. "Special Operations Command Is Looking at a New 6.5 Mm Round for Its Sniper Rifle." Military Times. Military Times, August 19, 2022. https://www.militarytimes.com/news/your-military/2017/04/18/special-operations-command-is-looking-at-a-new-6-5-mm-round-for-its-sniper-rifle/.

States., United. "Journals of the Continental Congress 1774-1789 / Edited from the Original Records in the Library of Congress V.2." HathiTrust. Accessed December 31, 2020. https://hdl.handle.net/2027/uc1.31822019445410?urlappend=%3Bseq.

Stevens, Charles Augustus. *Berdan's United States Sharpshooters in the Army of the Potomac, 1861-1865*. Bethesda, MD: University Publications of America, 1994.

Stewart, Herbert A. *From Mons to Loos: The Diary of a Supply Officer*. Auckland, NZ: Pickle Partners Publishing, 2013.

Stone, William L. *Campaign of Lieut. Gen. John Burgoyne: and the Expedition of Lieut. Col. Barry St. Leger. Hathitrust*. Albany, NY, 1877. .

Stossel, John, and Jordan B. Peterson. Jordan Peterson: The FULL Interview. Other, June 2018. .

Swofford, Anthony. *Jarhead: A Marine's Chronicle of the Gulf War and Other Battles*. New York City, NY: Scribner, 2003.

"The Famous Ukrainian Sniper 'Joan of Arc' Got Married in the Front-Line in the Kharkiv Region." *The Odessa Journal*, October 15, 2022. https://odessa-journal.com/the-famous-ukrainian-sniper-joan-of-arc-played-a-wedding-in-the-front-line-in-the-kharkiv-region/.

Thompson, Antonio Scott, and Christos G. Frentzos. *The Routledge Handbook of American Military and Diplomatic History: The Colonial Period to 1877*. London: Routledge, 2017.

Toland, John. *In Mortal Combat: Korea 1950 - 1953*. New York, NY: Open Road Integrated Media, 2016.

Troianovski, Anton, Valerie Hopkins, Ivan Nechepurenko, and Alina Lobzina. "Ukraine War Comes Home to Russians as Putin Imposes Draft." The New York Times. The New York Times, September 22, 2022. https://www.nytimes.com/2022/09/22/world/europe/putin-russia-military-ukraine-war.html.

Trepp, Caspar. "Letters from Captain Isler." *Papers*, September 24, 1862. https://bobcat.library.nyu.edu/permalink/f/22u4kq/nyu_aleph000849419.

Tzu, Sun. *The Art of War*. Translated by James Trapp. New York City, New York: Chartwell Books, 2012.

"Ultimate Guide on AGM-114 Hellfire Missiles: Capabilities and Cost." The Defense Post, March 30, 2021. https://www.thedefensepost.com/2021/03/22/agm-114-hellfire-missile/.

"United States Army Sniper Course: Course Description." Fort Benning | United States Army Sniper Course. United States Army. Accessed November 4, 2020. https://www.benning.army.mil/infantry/199th/Sniper/Description.html.

United States Department of Energy, A. W. Oughterson, G. V. Leroy, A. A. Liebow, E. C. Hammond, H. L. Barnett, J. D. Rosenbaum, and B. A. Schneider, 1 Medical Effects Of Atomic Bombs The Report Of The Joint Commission For The Investigation Of The Effects Of The Atomic Bomb In Japan § (1951).

United States Forces, Somalia after action report and historical overview: The United States Army in Somalia, 1992-1994 § (2003).

"U.S. Involvement in the Vietnam War: the Gulf of Tonkin and Escalation, 1964." U.S. Department of State. Accessed August 19, 2022. https://history.state.gov/milestones/1961-1968/gulf-of-tonkin.

Ward, Joseph T. *Dear Mom: A Sniper's Vietnam*. New York, NY: Presidio Press, 1991.

Washington, George. "Diary Entry: 10 February 1774." National Archives and Records Administration. Accessed November 3, 2020. https://founders.archives.gov/documents/Washington/01-03-02-0004-0003-0010.

———. "George Washington Papers, Series 3, Varick Transcripts, 1775-1785, Subseries 3G, General Orders, 1775-1783, Letterbook 2: Oct. 1, 1776 - Dec. 31, 1777." The

Library of Congress. Accessed November 3, 2020. https://www.loc.gov/item/mgw3g.002/.

———. "George Washington Papers, Series 3, Varick Transcripts, 1775-1785, Subseries 3B, Continental and State Military Personnel, 1775-1783, Letterbook 4: Aug. 1, 1777 - Jan. 20, 1778." The Library of Congress. Accessed November 3, 2020. https://www.loc.gov/item/mgw3b.004/.

"Whaling, William John 'Wild Bill,'" tracesofwar.com, accessed August 1, 2022, https://www.tracesofwar.com/persons/33816/Whaling-William-John-Wild-Bill.htm.

"William Whaling - Recipient." The Hall of Valor Project. Accessed August 1, 2022. https://valor.militarytimes.com/hero/8236.

Wright, Robert K. *The Continental Army*. Washington, D.C.: Center of Military History, U.S. Army, 1989. https://history.army.mil/html/books/060/60-4-1/CMH_Pub_60-4-1.pdf.

Zaĭt︠s︠ev V. G., and Neil Okrent. *Notes of a Russian Sniper: Vassili Zaitsev and the Battle of Stalingrad*. Barnsley, S. Yorkshire: Frontline Books, 2016.

www.ingramcontent.com/pod-product-compliance
Lightning Source LLC
Chambersburg PA
CBHW060435090426
42733CB00011B/2279